Reviews and critical articles covering the entire field of normal anatomy (cytology, histology, cyto- and histochemistry, electron microscopy, macroscopy, experimental morphology and embryology and comparative anatomy) are published in Advances in Anatomy, Embryology and Cell Biology. Papers dealing with anthropology and clinical morphology that aim to encourage cooperation between anatomy and related disciplines will also be accepted. Papers are normally commissioned. Original papers and communications may be submitted and will be considered for publication provided they meet the requirements of a review article and thus fit into the scope of "Advances". English language is preferred, but in exceptional cases French or German papers will be accepted.

It is a fundamental condition that submitted manuscripts have not been and will not simultaneously be submitted or published elsewhere. With the acceptance of a manuscript for publication, the publisher acquires full and exclusive copyright for all languages and countries.

Twenty-five copies of each paper are supplied free of charge.

Advances in Anatomy Embryology and Cell Biology

Vol. 170

Springer
Berlin
Heidelberg
New York
Barcelona
Hong Kong
London
Milan
Paris
Tokyo

Friedrich Paulsen

The Human Nasolacrimal Ducts

With 30 Figures and 8 Tables

Springer

PD Dr. med. Friedrich Paulsen
Institute of Anatomy
Christian-Albrechts-University of Kiel
Olshausenstr. 40
24098 Kiel
Germany

e-mail: fpaulsen@anat.uni-kiel.de

ISSN 0301-5556
ISBN 3-540-44076-3 Springer-Verlag Berlin Heidelberg New York

Library of Congress Cataloging-in-Publication Data
Paulsen, Friedrich, 1965
The human nasolacrimal ducts / Friedrich Paulsen.
p. cm. – (Advances in anatomy, embryology, and cell biology,
ISSN 0301-5556;.v.170
Includes bibliographical references and index.
ISBN 3540440763 (softcover: alk. paper)
1. Lacrimal apparatus. I. Title. II. Series.
QP188.T4 P38 2002 612.8'47–dc21

Springer-Verlag a member of BertelsmannSpringer
Science + Business Media GmbH

http://www.springer.de

Production: M. Uhing, Heidelberg
Typesetting, Stürtz, Würzburg

Printed on acid-free paper SPIN 10890295 27/3160MU – 5 4 3 2 1 0

Dedicated to my wife Dana

Contents

1	**Introduction and Questions**	1
1.1	Introduction	1
1.2	Questions	5
2	**Materials and Methods**	7
2.1	Human Material	7
2.1.1	Cadavers	7
2.1.2	Specimens Obtained at Surgery	7
2.1.3	Endonasal Dacryocystorhinostomy – Biopsy Technique	8
2.2	Animal Material	8
2.2.1	Cadavers	8
2.2.2	Living Animals	9
2.3	Light Microscopy	9
2.4	Lectin Staining	9
2.5	Immunohistochemistry	10
2.5.1	Extracellular Matrix	10
2.5.2	Trefoil Factor Peptides	11
2.5.3	Antimicrobial Peptides	11
2.5.4	Defence Cells	12
2.5.5	Nervous Tissue	12
2.6	Scanning Electron Microscopy	13
2.7	Transmission Electron Microscopy	13
2.8	Corrosion Vascular Casts	13
2.9	Reverse Transcription Polymerase Chain Reaction and Western Blot	14
2.9.1	Cell Culture	14
2.9.2	Total RNA Purification and cDNA Synthesis	14
2.9.3	RT-PCR of Trefoilfactor Peptides	14
2.9.4	Antisera and Western Blot Analysis of TFFs	15
2.9.5	RT-PCR of Antimicrobial Peptides	15
2.10	Animal Investigations/Experiments	16
2.10.1	Animal Experiments with Rats	16
2.10.1.1	Iodination of Bovine Serum Albumin	16
2.10.1.2	Absorption Experiments	16
2.10.1.3	Fractionation of Radioactivity in the Serum	16
2.10.1.4	Light Microscopy and Autoradiography	17
2.10.2	Comparative Anatomy	17

2.10.2.1 Ocular Sample Preparation and Light Microscopy for Comparative Anatomy 17
2.10.2.2 Scanning Electron Microscopy for Comparative Anatomy...................... 17
2.10.3 Animal Experiments with Rabbits 18
2.10.3.1 Determination of ^{3}H-Cortisol Uptake 18
2.10.3.2 Light Microscopy and Autoradiography......... 18

3 **Results** 19
3.1 The Epithelium and the Underlying Lamina Propria of the Nasolacrimal Ducts 19
3.1.1 Structure of the Epithelium..................... 19
3.1.1.1 Histology...................................... 19
3.1.1.2 Ultrastructure of the Epithelium................ 21
3.1.2 Receptor Binding of Carbohydrate-Specific Lectins to the Epithelium....................... 22
3.1.3 Trefoil Factor Peptides in the Epithelium........ 23
3.1.3.1 RT-PCR Analysis.............................. 23
3.1.3.2 Western Blot Analysis.......................... 23
3.1.3.3 General Histology and Immunohistochemistry.. 24
3.1.4 Localisation of Antimicrobial Peptides 27
3.1.4.1 RT-PCR Analysis.............................. 27
3.1.4.2 Antimicrobial Peptide Immunostaining......... 29
3.1.5 Mucosa Associated Lymphoid Tissue 31
3.1.5.1 Distribution and Organisation in "Healthy" (Cadavers) Nasolacrimal Systems............... 31
3.1.5.2 Distribution and Organisation in "Inflamed" (Surgically Obtained) Nasolacrimal Systems..... 33
3.1.5.3 Organisation of MALT in Cadavers and Surgically Obtained Specimens............. 35
3.1.5.4 IgA Immunostaining in "Healthy" (Cadavers) Nasolacrimal Systems 35
3.1.5.5 IgA Immunostaining in "Inflamed" (Surgically Obtained) Nasolacrimal Systems..... 35
3.2 The Membranous Part of the Nasolacrimal Ducts 38
3.2.1 The Vascular System 38
3.2.2 Innervation of the Vascular System and the Seromucous Glands of the Nasolacrimal Ducts .. 41
3.2.2.1 Distribution of Neuronal Markers 41
3.2.2.2 Ultrastructure of the Nervous Tissue........... 44
3.2.3 Structural Analysis of the Lamina Propria in the Nasolacrimal Ducts 45
3.3 Absorption Experiments in the Nasolacrimal Ducts 46
3.3.1 Absorption Experiments in Rats............... 46
3.3.1.1 Kinetics of Uptake of Radioactivity into the Serum 46
3.3.1.2 Analysis of the Incorporated Radioactivity in the Serum................................. 46

3.3.1.3 Light Microscopy and Autoradiography 46
3.3.2 Comparative Anatomy of the Nasolacrimal Ducts 50
3.3.2.1 Comparative Histology 50
3.3.2.2 Comparative Scanning Electron Microscopy (Human and Rabbit) 52
3.3.3 Absorption Experiments in Rabbits 52
3.3.3.1 Kinetics of Uptake of Radioactive Cortisol into Blood/Serum 52
3.3.3.2 Light Microscopy and Autoradiography (Rabbit) 53
3.4 Pathologic Findings in Dacryostenosis 53
3.4.1 Histological Findings 55
3.4.2 Ultrastructural Findings 57

4 Discussion 59
4.1 Mechanisms of Immune Defence in the Nasolacrimal Ducts 59
4.1.1 The Function of the Epithelium Lining the Nasolacrimal Ducts 59
4.1.1.1 Lectin Binding Sites and Mucins 59
4.1.1.2 TFF Peptides 61
4.1.1.3 Antimicrobial Peptides 63
4.1.2 Specific Immune Defence by MALT 66
4.1.2.1 MALT of the Nasolacrimal Ducts 66
4.1.2.2 Loss of TALT in Association with the Scarring of Symptomatic Dacryostenosis 68
4.2 Mechanisms Regulating Tear Flow Through the Nasolacrimal Ducts 70
4.2.1 Regulation of Tear Flow by the Cavernous Body . 70
4.2.2 Innervation of the Cavernous Body and the Seromucous Glands 72
4.2.3 Experimental Influence on the Regulation of Tear Flow by the Cavernous Body 74
4.2.4 Biomechanics of Tear Flow 74
4.3 Absorption of Tear Fluid Components........... 75
4.3.1 What Happens to Tears Inside the Nasolacrimal Ducts? 75
4.3.2 The Nasolacrimal System of the Rabbit Shows High Homology with the Human Nasolacrimal Ducts ... 77
4.3.3 The Rabbit – Animal Model for the Absorption of Lipophilic Substances from Tear Fluid by the Epithelium of the Nasolacrimal Ducts 77
4.4 Insights into the Pathophysiology of Primary Acquired Dacryostenosis 79
4.5 Considerations on the Relationship Between Dry Eye and the Nasolacrimal Ducts 82
4.5.1 Absorption of Tear Fluid in the Nasolacrimal Ducts – Feedback Signal for Tear Fluid Production?................................. 82

4.5.2 Tear Duct-Associated Lymphoid Tissue – Entrance Side for the Immunological Events in Dry Eye? 83

5 **Outlook** 87

6 **Summary**.. 89

References 91

Subject Index 105

Acknowledgements

Special thanks are due to my mentor *Bernhard Tillmann* as well as *Andreas Thale* and *Ulrich Schaudig* as strength of the '**Nasolacrimal Duct Research Group**'.

My grateful thanks are also extended to the following persons named in alphabetical order who contributed to the present investigations
Muhammad Ajub, Monica Berry, Anthony Corfield, Marc Föge, Uta Hallmann, Janka Held-Feindt, Margitta Hinz, Roman Hirt, Werner Hoffmann, Guido Kohla, Jan Lehmann, Steffen Maune, Rolf Mentlein, Reza Parwaresch, Jens Paulsen, Thomas Pufe, Rainer Rochels, Roland Schauer, Jens-Michael Schröder, Philipp Steven.

Moreover, I would like to thank *Frank Lichte, Sonja Reiter, Karin Stengel* and *Regine Worm* for their excellent technical assistance; *Heidi Waluk* and *Heidi Siebke* for photographic work; *Clemens Franke* for the schematic graphs; *Rolf Klaws* for excellent support with the corrosion vascular casts; and *Michael Beall* for editing the English.

These studies were supported by Deutsche Forschungsgemeinschaft grants Pa 738/1-1, Pa 738/1-2 and Pa 738/1-3, programs of the German Research Foundation.

Finally, my particular thanks are meant for the patients who gave permission for their tissues that were removed at surgery to be investigated as well as the persons who dedicated their body to the Institute of Anatomy, Christian-Albrechts-University of Kiel, Germany by testament and thus enabled the present investigations.

1 Introduction and Questions

1.1 Introduction

Tear fluid is drained by the nasolacrimal ducts (Fig. 1a, b) into the inferior meatus of the nose. The first exact description of the nasolacrimal ducts originates form Giovanni Battista Carcano Leone (1574). His explanations lead, together with the work of Niels Stensen (1662) on tear secretion, to a plausible anatomical conception of the entire lacrimal system.

The lacrimal passages can be divided into a bony passage and a membranous lacrimal passage (Duke-Elder 1961). The bony passage is formed anteriorly by the frontal process of the maxilla and posteriorly by the lacrimal bone. The membranous lacrimal passages include the lacrimal canaliculi, the lacrimal sac and the nasolacrimal duct.

The upper and lower canaliculus are lined by a pseudostratified/stratified columnar epithelium and are surrounded by a dense ring of connective tissue as well as by muscle fibres of the lacrimal portion of the orbicularis oculi muscle (Horner's muscle; Halben 1904). The lacrimal sac and the nasolacrimal duct are lined by a double-layered epithelium revealing a superficial columnar layer and a deep flattened layer of basal cells (Tsuda 1952; Duke-Elder 1961). Both layers sometimes appear as a pseudo-stratified epithelium. Radnot and Bölcs (1971) and Radnot (1977) describe kinocilia in the nasolacrimal ducts to be a common finding but only to occur in very low numbers. Beside epithelial cells, goblet cells are integrated in the epithelium sometimes forming intraepithelial mucous glands (Werncke 1905). Moreover, small serous glands are present in the lamina propria, especially in the fundus of the lacrimal sac (Rivas et al. 1991).

The preocular tear film is composed of three components, a lipid, aqueous and mucous component. The three parts are produced by different structures. The lipid component is secreted by large oil glands in the eyelids called Meibomian glands, the aqueous tear component by the lacrimal gland as well as the accessory lacrimal glands (glands of Krause; glands of Wolfring) of the lids, and the mucous by goblet and epithelial cells of the conjunctiva. It is unknown whether the normal tear film is retained or whether there occurs a change of tear fluid composition inside the nasolacrimal ducts.

A bulk of work has done on the mucus part of the preocular tear film, characterising its single components. This mucus part, consisting primarily of soluble mucin glycoproteins, serves to lubricate and protect the superficial ocular epithelium, to retard fluid evaporation, and to anchor the tear film to the ocular surface (Lemp 1973;

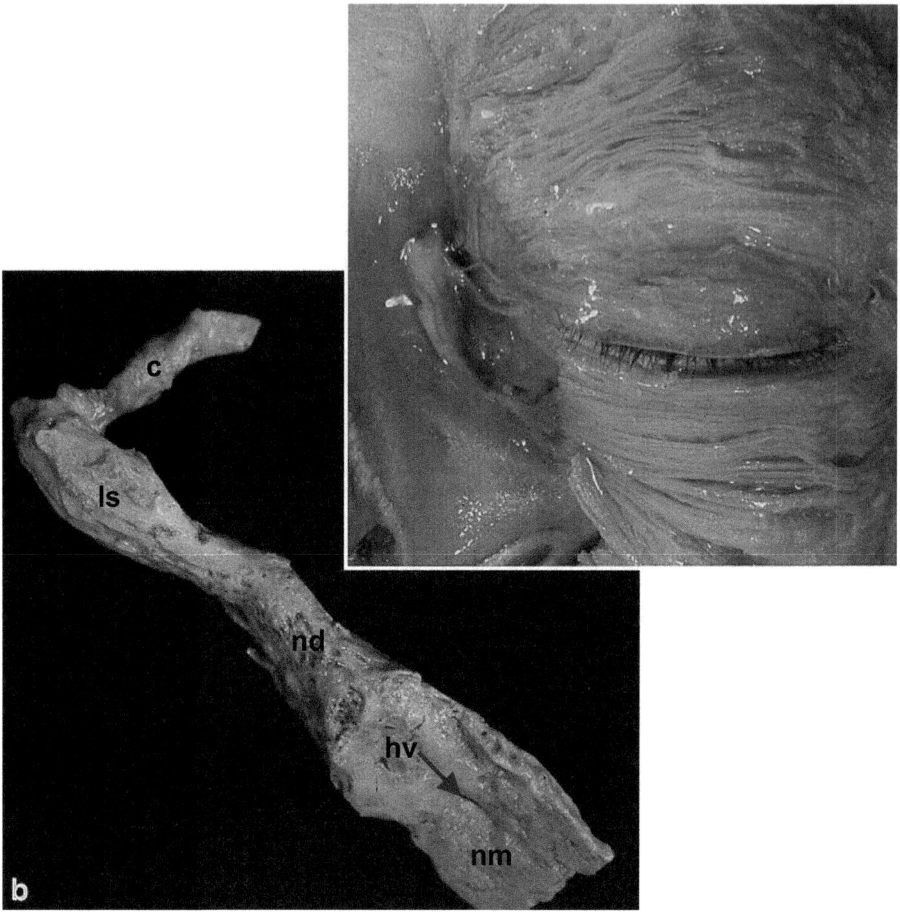

Fig. 1a, b. Macroscopy of the nasolacrimal ducts. **a** View of a prepared lacrimal system in situ. **b** Macroscopic view of a prepared efferent tear duct system removed from its bony channel. *c*, Lacrimal canaliculi – individual canaliculi not distinguishable; *ls*, lacrimal sac, *nd*, nasolacrimal duct; *nm*, mucous membrane of the nose; *hv*, Hasner's valve opening of the nasolacrimal duct into inferior meatus of the nose

Dilly 1994). Recent studies indicate that the hydrated mucus occupies 70%–90% of the tear film thickness, suggesting that soluble mucins are an important structural component of the precorneal tear film (Chen et al. 1997). Also, mucins function as buffers, antioxidants and inhibitors of bacterial adhesion (Gendler and Spicer 1995). Moreover, they are suggested to influence the rheological properties of the ocular mucus (Corfield et al. 1997; Gipson and Inatomi 1997). The rheological properties are defined by the tear break-up time, which is changed in various pathologic conditions (e.g. in patients with dry eye symptoms). In this context it has been shown that goblet cells of the porcine and human conjunctiva possess trefoil-factor peptides (TFF peptides, formerly P-domain peptides, trefoil factors; Jagla et al. 1999; Langer et al. 1999) which, like mucins, are typical constituents of mucus gels. The physiological functions of TFF peptides are multiple. They promote migration of epithelial cells in vitro (Dignass et al. 1994), enhance mucosal healing and epithelial restitution in vivo in the gastrointestinal mucosa (Kindon et al. 1995; Alison et al. 1995), have anti-apoptotic properties (Chen et al. 2000; Taupin et al. 2000) and can protect the gastrointestinal epithelium from a variety of noxious agents, including bacterial toxins, chemicals and drugs (Kindon et al. 1995, Playford et al. 1995; Babyatsky et al. 1996; Mashimo et al. 1996; Taupin et al. 2000). Moreover, synthesis of TFF peptides has been shown to occur pathologically as result of chronic inflammatory diseases, and aberrant synthesis of TFF peptides has been observed in many tumours (Hoffmann et al. 2001). TFF peptides are also suggested to have a great impact on the rheological properties of the tear film (Langer et al. 1999). Nearly nothing is known about mucins and TFF peptides of the nasolacrimal ducts.

Inflammatory processes in the area of the nasolacrimal ducts often show pathological changes in their follow-up with functional restriction of tear flow. The pathomechanisms on the cellular level, especially the role of the lining epithelium of the nasolacrimal ducts in immune response, is largely unclear.

During recent years, endogenous antimicrobial peptides, active against gram-positive and gram-negative bacteria, have attracted much interest in the ocular system as well (Cullor et al. 1990; Schwab et al. 1992; Gottsch et al. 1998; Haynes et al. 1998, 1999, 2000; McNamara et al. 1999; Lehman et al. 2000; Zasloff 2002). Moreover, some studies suggest that invertebrate epithelial cells can recognise microorganisms and mount a rapid defence response through the production of various inducible antibiotic peptides as shown for *Drosophila* (Tzou et al. 2000). This leads to a characteristic broad spectrum of antimicrobial activity against bacteria and fungi (Bowman 1995). It is unknown whether the epithelial cells of the nasolacrimal ducts are also able to release antimicrobial peptides, especially in the presence of inflammation like dacryocystitis.

In the context of specific immunity, Marcovitch (1951) and Tsuda (1952) recognised the frequent occurrence of intraepithelial lymphocytes in the epithelium of the nasolacrimal ducts. Perra et al. (1995) show high amounts of IgA inside and on the surface of the lining epithelium of the nasolacrimal ducts. Duke-Elder (1961) described underneath the epithelium a layer of adenoid tissue, sometimes aggregated into follicles.

The evidence on primary extranodal marginal zone B-cell lymphoma [low-grade B-cell lymphoma of the mucosa-associated lymphoid tissue (MALT) type] of the efferent tear ducts (Kheterpal et al. 1994; White et al. 1995; Tucker et al. 1997) infers the presence of pre-existent organised tissue at that site. It is not clear whether organised MALT is a normal component of the human efferent lacrimal system.

Tear flow from the lacrimal sac to the cavum nasi is caused by forces that are not completely understood. Also, the function of the drainage structures regarding tear fluid composition is still unknown. Many factors involved in tear outflow have been hypothesised, dependent on the unique anatomical configuration of the nasolacrimal passages. An active mechanism has been recognised as an essential factor in lacrimal drainage, since the observation of epiphora in cases of facial palsy (von Arlt 1855). This mechanism was investigated by Frieberg (1917) and Rosengreen (1928) who found evidence for a canalicular pump. The concept of a canalicular pump was supported by anatomic studies (Rohen 1953; Jones 1961), high-speed photography (Doane1981), scintillography (Chavis et al. 1978) and intracanalicular pressure measurements (Hill et al. 1974; Wilson and Merril 1976). Some hypotheses have also assumed an active pump mechanism to explain the function of the human lacrimal sac (Jones 1958; 1961; Nagashima and Araki 1963; Becker 1992). Rohen suggests reabsorption of tear fluid by the lining epithelium of the nasolacrimal ducts. Moreover, physical factors such as evaporation (Schirmer 1903), gravity (Murube del Castillo 1978), respiration (Nik et al. 1984) and capillarity (Hurwitz 1996) have been suggested as possible factors in tear outflow.

Although the physiology of lacrimal drainage has been under study for more than a century, the pathophysiology of functional lacrimal drainage insufficiency is still not understood (i.e. cases of epiphora despite patent lacrimal passages found when syringing).

As early as in 1866, Henle described a vascular plexus surrounding the lumen of the lacrimal sac and the nasolacrimal duct. This network of large vessels is connected caudally with the cavernous body of the nasal inferior turbinate (Duke-Elder 1961; Rohen 1964). Although more than two-thirds of the bony canal between orbit and inferior turbinate is occupied by this wide-lumened vascular plexus, textbooks of anatomy do not mention its existence.

Epiphora resulting from obstruction of the nasolacrimal ducts is a common problem for the ophthalmologist and the otorhinolaryngologist, both of which are integrated in the diagnostic and therapeutic process in most cases. Diagnosis and treatment of this condition require a thorough understanding of the lacrimal apparatus and its ocular and nasal relationships. Idiopathic or primary acquired dacryostenosis, synonymous with primary acquired nasolacrimal duct obstruction (PANDO), is a syndrome of unknown aetiology. Of all nontraumatic forms, it accounts for the vast majority of cases observed in adults. Pathologic studies of the lacrimal passage have indicated that primary acquired dacryostenosis results from fibrous obstruction secondary to chronic inflammation (Busse and Müller 1977; Linberg and McCormick 1986; Mauriello et al. 1992). Nevertheless, the pathophysiology of functional dacryostenosis, i.e. patients with epiphora in spite of patent lacrimal passages on syringing, still has not been understood. The pathogenesis has also been controversial. Descending pathogens from the conjunctival sac (Summerskill 1956) as well as diverticula of the lacrimal passage (Melanová 1969) are supposedly causal. Other specialists claim to have located the origin of idiopathic dacryostenosis in the nose (Traquair 1941; Dorello 1957; Köhler and Müller 1968; Müller et al. 1968). Here, simple infections of the nasal mucous membrane (Busse and Müller 1977) or diseases of the sinuses (Traquair 1941; Dorello 1957) have been suggested.

According to a newer classification, "dry eye" (synonymously termed sicca syndrome or keratoconjunctivitis sicca) is defined as a disorder of the tear film due to tear deficiency or excessive evaporation that causes damage to the interpalpebral ocular surface and is associated with symptoms of discomfort (Lemp 1995). This common definition includes all forms of the sicca syndrome, idiopathic as well as the forms that are associated with systemic disease (like Sjögrens syndrome). The role of keratoconjunctivitis sicca has demonstrably increased in Germany and has become one of the most common ocular diseases. Although knowledge in the pathogenesis, diagnosis and treatment of dry eye and tear film disorders has made great advances during the last 25 years, there are still many questions unanswered. In this context, the question has never been raised whether the nasolacrimal ducts may play a role in dry eye.

1.2 Questions

The aim of the present study was to analyse the human nasolacrimal ducts under structural, functional, as well as immunological aspects and to get deeper insight into the pathophysiology of diseases concerning the nasolacrimal system. This was conducted by means of histology, histochemistry, immunohistochemistry, scanning electron microscopy, transmission electron microscopy, molecular biology and maceration techniques, as well as by corrosion vascular casts, animal and human experiments. The investigations try to give answers regarding the following questions:

1. Which mechanisms act against inflammation in the nasolacrimal ducts?
2. Which mechanisms support tear drainage through the nasolacrimal ducts?
3. Are tear components reabsorbed in the nasolacrimal system?
4. What is the pathogenesis of functional dacryostenosis, i.e. patients with epiphora in spite of patent lacrimal passages on syringing?
5. Do the nasolacrimal ducts have a role in the pathogenesis of dry eye?

Results may open new perspectives compared with existing therapeutic concepts against diseases of the nasolacrimal system.

2 Materials and Methods

2.1 Human Material

2.1.1 Cadavers

Heads of adults and lacrimal systems were obtained from cadavers donated to the Institute of Anatomy, Christian-Albrechts University of Kiel, Germany, and were prepared for histological, histochemical, enzyme-histochemical, immunohistochemical, molecular biological, scanning electron microscope and transmission electron microscope investigations as well as for vascular corrosion casts. Limited information was available on the specimens; however, the specimens were obtained from individuals free of recent trauma, eye or nasal infections, and diseases potentially involving or affecting lacrimal function. Moreover, lacrimal systems obtained during surgical procedures (i.e. dacryostenosis, dacryocystitis, orbital exenteration, tumour dissection of the paranasal sinuses with resection of the lateral nasal wall) were prepared for the same methods as already described above. Special groups of patients were selected for studies on antimicrobial peptides (Tables 3 and 4) and the pathophysiology of primary acquired dacryostenosis (Table 6). All material originated from the Department of Otorhinolaryngology, Head and Neck Surgery, University of Kiel, Kiel, Germany and from the Department of Ophthalmology, University Hospital Hamburg-Eppendorf, Hamburg, Germany, and was obtained with the permission of the medical ethics commission of the University of Kiel and used in accordance with the Declaration of Helsinki. Except for the size of the removed lacrimal systems, there were no individual differences or differences in freshly obtained material versus the material obtained from fixed bodies.

2.1.2 Specimens Obtained at Surgery

Material from surgical procedures was obtained with the permission of the Medical Ethics Committee and used in accordance with the Declaration of Helsinki. Patients were selected from referrals to the Department of Otorhinolaryngology, Head and Neck Surgery, Christian-Albrechts University of Kiel, Germany between March 1997 and March 2001. Patients presenting with epiphora were carefully interviewed with regard to the duration of tearing, history of acute or chronic dacryocystitis and complaints of mucoid discharge in the medial canthal region. Following anamnestic exclusion of facial surgery or trauma, allergies, family history of tearing, external eyelid disease and topical eye medication, eyelid malposition and periocular neoplasm were excluded by a careful ocular examination at the Department of Ophthalmology, Christian-Albrechts University of Kiel, Germany. The lacrimal passage was assessed by inspection, palpation, digital expression of lacrimal sac contents, Jones testing, syringing of the upper and lower canaliculus, dacryocystography and axial and coronary computer tomography in selected cases. Patients revealing punctal or canalicular stenosis were excluded from the study. Moreover, a nasal endoscopic examination was performed in each case at the Department of Otorhinolaryngology, Head and Neck Surgery, Christian-Albrechts University.

No microbiologic tests were performed on the tearway fluid of body donors or on that of patients prior to processing for histology.

2.1.3 Endonasal Dacryocystorhinostomy – Biopsy Technique

Endonasal dacryocystorhinostomy was performed using a modified version of the conventional method according to West (1909). Briefly, after resection of the bony walls of the medial part of the frontal process of the maxilla and of the os lacrimale, the periosteum of the fossa lacrimalis and the ductus nasolacrimalis was visible. By partly removing it, a broad entrance was opened into the lacrimal fossa and the nasolacrimal canal. From the lower canaliculus, the lacrimal system was penetrated by pushing a stylet forward to the stenotic area. The tissue surrounding the lumen of the lacrimal sac above the stenosis was then opened by scalpel cutting to the stylet. Parts of the incision margins were removed with a small forceps and a broad exit of the lacrimal sac into the nose was produced. It was then possible to remove the whole stenotic area below this exit by cutting the whole lacrimal system horizontally above and below the stenotic area and extirpating it. Endonasal procedures were performed under endoscopic or microscopic view without destroying important adnexal structures.

2.2 Animal Material

2.2.1 Cadavers

All of the animals used for the comparative studies had already died. Heads of rabbits and deer were obtained from hunters. Heads of apes (*Macaca mulatta*) were obtained from the German Primate Centre, Göttingen, Germany. Heads of rats were obtained from the animal house of the Preclinical Institutes, University of Kiel, Germany, and heads of cats from the Institute of Physiology, University of Kiel, Germany. Heads of pigs were obtained from a local slaughterhouse.

2.2.2
Living Animals

Living Wistar rats and Chinchilla bastard rabbits were used for the absorption experiments. The tests were permitted by the Ministry of Environment, Nature and Forests, Schleswig-Holstein, Germany (permit numbers X 330a-72241.121-1 [6-1/98] and X330a-72241.121-1 [74-7/00]) and were carried out in accordance with guidelines for care and use of laboratory animals.

2.3
Light Microscopy

For analysis by light microscopy, tissue specimens obtained during surgical procedures and from cadavers as well as from animal cadavers were fixed in 4% formalin, decalcified in 20% ethylenediaminetetraacetate (EDTA) as necessary, dehydrated in graded concentrations of ethanol and embedded in paraffin. Frontal, sagittal and horizontal sections (7 μm) were deparaffinised and stained with toluidine blue (pH 8.5), haemalum, Sudan III haemalum, azan, resorcin-fuchsine-thiazine picric acid, orcein, alcian blue (pH 1) and using Goldner and Gomori staining according to instructions of Romeis (1989) as well as by mild periodic acid–Schiff (mPAS) staining according to instructions of Veh et al. (1979). The slides were examined by microscope (Axiophot; Zeiss, Oberkochen, Germany).

2.4
Lectin Staining

For lectin-binding experiments (Sharon and Lis 1972), 7-μm-thick tissue sections were deparaffinated in xylene, rehydrated with a graded series of ethanol, passed into aqua bidest and finally into Tris-buffered saline (TBS) (0.05 M Tris, 0.15 M sodium chloride, pH 7.4), supplemented with 0.02 M $CaCl_2$. Incubation of the sections with various fluorescein isothiocyanate (FITC)-conjugated lectin solutions (10 μg/ml) was performed in a dark moist chamber at room temperature for 30 min. After washing the slides three times with TBS, sections were mounted in 10% glycerol in TBS or in Vectashield (Vecto Laboratories, Burlingame, Calif., USA). The specificity of the lectin binding was checked in control experiments by mixing the lectin solution with the corresponding inhibiting sugar (Table 1) at a concentration of 0.2–0.5 M prior to incubation for 30 min at room temperature. The origin and the specificities of the lectin conjugates are given in Table 1. To estimate the autofluorescence of the tissue, TBS-mounted sections were used. Slides were examined using a Zeiss-Axiophot microscope equipped for epifluorescence.

Table 1. Origin and carbohydrate specificities of lectin conjugates

Lectin	Abbre-viation	Carbohydrate specificities	Inhibitor
Concanavalin A agglutinin	ConA	α-Man	α-methyl-Man+α-methyl Glc
Griffonia simplicifolia I agglutinin	GSA-I	α-GalNAc>α-Gal	GalNAc+Gal
Griffonia simplicifolia II agglutinin	GSA-II	α(β)-GlcNAc	GlcNAc
Limulus polyphemus agglutinin	LPA	Neu5Acα(2-3)GalNAc>Neu5Acα(2-6)GalNAc	BSM
Maackia amurensis agglutinin	MAA	Neu5Acα(2-3)Gal	Neu5Ac
Sambucus nigra agglutinin	SNA	Neu5Acα(2-6)Gal	Neu5Ac
Ulex europaeus I agglutinin	UEA-I	α-L-Fuc	Fuc
Wheat germ agglutinin	WGA	α-GlcNAc>α-GalNAc>Neu5Ac	GlcNAc, Neu5Ac
Succinylated wheat germ agglutinin	sucWGA	α-GlcNAc>α-GalNAc	GlcNAc
Maclura pomifera agglutinin	MPA	α-GalNAc	GalNAc

BSM, bovine submandibular gland mucin; Fuc, fucose; Gal, galactose; GalNAc, *N*-acetyl-galactosamine; Glc, glucose; GlcNAc, *N*-acetyl-glucosamine; Man, mannose; Neu5Ac, *N*-acetylneuraminic acid.

2.5 Immunohistochemistry

In all cases, the efferent tear duct systems were dissected and freed from the surrounding bone before embedding in paraffin or freezing in liquid nitrogen.

2.5.1 Extracellular Matrix

Immunohistochemical investigations were performed on unfixed material, which was snap frozen in liquid nitrogen. Sagittal and horizontal sections (7 μm) were cut with a cryostat at –21°C and mounted on gelatine-coated slides. For immunohistochemistry, frozen sections were pretreated with testicular hyaluronidase (Boehringer, Mannheim, Germany) in TBS in a moist chamber at 37°C for 30 min. The sections were washed three times with TBS and incubated with goat serum for 45 min at room temperature. Incubation with the primary antibody was carried out for 60 min at room temperature. The following antibodies were used: mouse anti-collagen (MAC) type I polyclonal antibody (1:10; Bio-Science Products AG, Emmenbrücke, Switzerland), MAC type III polyclonal antibody (1:10; Bio-Science Products AG, Emmenbrücke, Switzerland), anti-chondroitine-4-sulphate monoclonal antibody (1:100; Chemicon International, Temecula, Calif., USA), anti-chondroitine-6-sulphate monoclonal antibody (1:100; Chemicon International, Temecula, Calif., USA), and anti-keratan sulphate monoclonal antibody (1:500; Bio-Science Products AG, Emmenbrücke, Switzerland). The presence of basement membrane components was investigated using an antibody against laminin (1:50; Medac, Hamburg, Germany). Sections were labelled with the respective secondary antibody, FITC – conjugated goat anti-rabbit or goat anti-mouse IgG for 45 min. Control sections were incubated only with the FITC-conjugated antibody. Tissues with defined antigen sites (human cartilage, skin, kidney, liver, spleen) were used as positive controls. The slides were examined with a Zeiss-Axiophot microscope equipped for epifluorescence.

2.5.2 Trefoil Factor Peptides

Immunohistochemical stains were done with antibodies against TFF1 (anti-TFF1, 1:5,000 dilution) and TFF3 (anti-rTFF3-1, 1:1,000 dilution). They were applied using a standard peroxidase-labelled streptavidin-biotin technique, using conventional methods with trypsinisation. After counterstaining with haemalum, the sections were finally mounted with Aquatex (Boehringer, Mannheim, Germany). For co-localisation studies of TFF1 and TFF3, the EnVision Doublestaining system (Dako, Glostrup, Denmark) was used. Sections were first stained with anti-TFF1, revealing a brown reaction product, and afterwards with anti-rTFF3-1, revealing a red reaction product. Two negative control sections were used in each case. One was incubated with the second antibody only, the other with the primary antibody only. Sections of human submandibular gland and human jejunal mucosa were used for positive control. Furthermore, specificity of the TFF1 and TFF3 staining was tested by competition with the corresponding synthetic peptide, i.e. 1 ml TFF1 blocking peptide (1:500 dilution, Santa Cruz Biotechnology, Santa Cruz, CA) was pre-absorbed with 10 μg human TFF1 for 16 h at 4°C and 1 ml anti-rTFF3-1 (1:5,000 dilution) was pre-absorbed with 10 μg synthetic FKPLQEAECTF (representing the c-terminus of human TFF3) for 16 h at 4°C and then used for immunohistochemistry. All slides were examined by microscope (Zeiss-Axiophot). Photomicrographs were taken on Kodak Ektachrome 64.

2.5.3 Antimicrobial Peptides

Immunohistochemical stains were performed with antibodies against lysozyme (1:200 in TBS, 60 min; A0099; Dako, Glostrup, Denmark), lactoferrin (1:150 in TBS, 60 min; A0186, Dako), secretory phospholipase A_2 (sPA_2, 1:10 in TBS, 60 min; 05–143 mouse monoclonal sPA_2, Upstate Biotechnology, Lake Placid, N.Y., USA), human neutrophil defensins (HNP-1–3, 1:800 in TBS, 60 min, T-1034 mouse monoclonal Def-3, Bachem, Heidelberg, Germany), human β-defensin (HBD)-1, and HBD-2 (both 1:500 in TBS, 60 min; courtesy of Prof. T. Ganz, Pulmonary Research Laboratory, Los Angeles, CA, USA). They were applied using a standard peroxidase-labelled streptavidin-biotin technique, either with microwave heating pretreatment or using conventional methods with trypsinisation where appropriate. After counterstaining with haemalum, the sections were finally mounted with Aquatex (Boehringer, Mannheim, Germany). Two negative control sections were used in each case. One was incubated with the second antibody only, the other with the primary antibody only.

Sections of human submandibular gland (lysozyme, lactoferrin), human jejunal mucosa (phospholipase A_2), human spleen (Def-3) and human skin (HBD-1) were used for positive control. All slides were examined by microscope (Zeiss-Axiophot). For HBD-2, normal human skin was used as an additional negative control, since a positive control was not available.

2.5.4 Defence Cells

Immunohistochemical staining was performed using antibodies against CD20 (L26; concentrated), CD3 (1:100), immunoglobulin (Ig) M (1:100), IgA (1:20), IgG (1:50), follicular dendritic cells (KiM4; concentrated), CD45RA (KiB3; concentrated) and CD68 (KiM1; concentrated). With the exception of Ki, all antibodies were from Dako (Glostrup, Denmark); the Ki antibodies were provided by the Department of Pathology, Christian-Albrechts University of Kiel, Germany. They were applied using a standard peroxidase-labelled streptavidin-biotin technique (Bains and Miller 1988), either with microwave heating pretreatment (Cattoretti et al. 1993) or using conventional methods with trypsinisation where appropriate. Two negative control sections were used in each case. One was incubated only with the second antibody, the other only with the primary antibody. As a positive control, sections of human spleen (all other antibodies) were used. All slides were examined using a microscope (Zeiss-Axiophot).

2.5.5 Nervous Tissue

For immunohistochemical analysis, samples were fixed in Zamboni's solution (Zamboni and de Martino 1967) overnight at 4°C, then rinsed several times in 0.1 M phosphate-buffered saline (PBS) and saturated overnight at 4°C with 10% and 30% sucrose in 0.1 M phosphate buffer. This was followed by freezing and sectioning (7 μm) in a cryostat. After preincubation with 10% standard goat serum in PBS, the sections were incubated with various primary antisera to human neuropeptides and neuronal markers in a humidity chamber at room temperature. The following antibodies were used: mouse anti-reverse transcriptase (RT) 97 monoclonal antibody (1:200; Boehringer, Mannheim, Germany); mouse anti-S-100 monoclonal antibody (concentrated; Boehringer); rabbit anti-vasoinhibitory peptide (VIP) polyclonal antibody (1:800; Biotrend, Köln, Germany); rabbit anti-CGRP polyclonal antibody (1:100, Boehringer); rabbit anti-neuropeptide Y (NPY) polyclonal antibody (1:2,500; Cambridge Research Chemicals, Cambridge, UK); rabbit anti-tyrosine hydroxylase (TH) monoclonal antibody (1:800; Chemicon International, Temecula, Calif., USA); mouse anti-neuron-specific enolase (NSE) polyclonal antibody IgG (1:100; Dako, Glostrup, Denmark); rabbit anti-signal peptide (SP) polyclonal antibody (1:2000; Biotrend). Following careful rinsing in PBS, the secondary antibody was inoculated (30 min with biotin-labelled species-specific secondary antibodies, 1:200; Dako, diluted in inactivated human serum, 1:20 with PBS). Sections were then incubated for 30 min with avidin biotin complex (ABC) (1:100; Dako). Following the peroxidase-substrate solution reaction (Sigma, Deisenhofen, Germany), sections were rinsed with cold tap water, counterstained with haemalum and, finally, mounted with DePeX. For immunohistochemistry of S-100, a FITC-conjugated goat anti-mouse IgG was used for 45 min as secondary antibody: two negative control sections were used in each case. One was incubated with the second antibody only, the other with the primary antibody only. Sections of human peripheral nerve and muscle were used as positive controls. All slides were examined under a Zeiss Axiophot microscope also equipped for epifluorescence.

2.6 Scanning Electron Microscopy

For scanning electron microscopy, samples obtained from cadavers were fixed in 2.5% glutaraldehyde for 1 week. Some of them were cut longitudinally and opened into halves to examine the lining epithelium. Others were macerated with HClO (Paulsen and Thale 1998) or NaOH (Ohtani et al. 1988) to remove cellular compartments. All tissue blocks were then saturated in 2.5% tannic acid for 2 days. A counter fixation in 2% OsO_4 for 4 h was followed by dehydration in ethanol and drying in a critical point dryer. Preparations were coated with gold and analysed with a Philips scanning electron microscope.

2.7 Transmission Electron Microscopy

Transmission electron microscopy (TEM) samples obtained from cadavers and during endonasal surgery were fixed in 3.5% glutaraldehyde (in 0.1 M Sørensen phosphate buffer solution at pH 7.4) at 4°C for 1 week. Following dehydration in graded concentrations of ethanol, the samples were embedded in araldite. Semi-thin sections were prepared for light microscopy and stained with toluidine blue (pH 8.5); ultrathin sections were contrasted with uranyl acetate and lead citrate. Examination was done with a Zeiss (Oberkochen, Germany) electron microscope (TEM 902).

2.8 Corrosion Vascular Casts

For corrosion vascular casts, the bony cranium and brain were removed from heads of donor cadavers. The right and left ophthalmic arteries were exposed. Cannulae were introduced into these arteries, and the external and internal carotid arteries were ligated. Ten millilitres of casting resin mixture (Mercox CL-2B and MA, Dainihon Ink Chemical, Tokyo, Japan; or Acrifix 190+ Katalysator 20, Fa.; Röhm, Darmstadt, Germany) were injected through each cannula under hand pressure. After the resin had polymerised, the eyes were removed from the orbit, and the heads were transferred to plastic containers. To obtain completely macerated specimens, the organic material was removed with 5% potassium hydroxide solution and maintenance of a maceration temperature of 40°C. Maceration was then continued in distilled water at a temperature of 40°C. Finally, the specimens were air dried. In two heads the vascular system of the efferent lacrimal ducts were removed from the bony canal, sputter coated and viewed under the scanning electron microscope.

2.9 Reverse Transcription Polymerase Chain Reaction and Western Blot

2.9.1 Cell Culture

The human epithelial cell line Hakard (Boukamp et al. 1988; as a positive control for HBD-1) was obtained from the Department of Dermatology, Christian-Albrechts University of Kiel, Germany, and cultivated as well as subcultivated (after trypsinisation) in Dulbecco's modified Eagle's medium (DMEM) containing 10% fetal calf serum.

2.9.2 Total RNA Purification and cDNA Synthesis

For RT-PCR, frozen samples (20 mg) were crushed in an agate mortar under liquid nitrogen, the RNA was isolated by the phenol-guanidinium thiocyanate method, purified by isopropanol and repeated ethanol precipitation, and contaminating DNA was destroyed by digestion with RNase-free DNase I (20 min 25°C; Boehringer, Mannheim, Germany). After inactivation of the DNase (15 min 65°C), cDNA was generated with 1 µl (20 pmol) oligo $(dT)_{15}$ primer (Amersham Pharmacia Biotech, Uppsala, Sweden) and 0.8 µl superscript RNase H-reverse transcriptase (Gibco, Paisley, UK) for 60 min at 37°C as described by Feindt et al. (1995).

RNA was extracted from the culture cell lysates with the Rneasy Total RNA Kit (Qiagen, Chatsworth, Calif., USA) as per manufacturer's instructions. cDNA was then generated with 1 µl (20 pmol) oligo $(dT)_{15}$ primer (Amersham Pharmacia Biotech, Uppsala, Sweden) and 0.8 µl superscript RNase H-reverse transcriptase (Gipco, Paisley, UK) for 60 min at 37°C (Feindt et al. 1995).

2.9.3 RT-PCR of Trefoilfactor Peptides

For RT-PCR, frozen samples (20 mg) of all 18 patients and of four body donors were crushed in an agate mortar under liquid nitrogen and the RNA was isolated using a guanidium thiocyanate protocol. RNA purification via CsCl ultracentrifugation and RT-PCR analysis monitoring expression of TFF1, TFF2 and TFF3 were essentially as described previously (Wiede et al. 2001), with 30 amplification cycles (Taq DNA polymerase; Quiagen GmbH, Hilden, Germany). As a control, glyceraldehyde-3-

phosphate dehydrogenase (GAPDH) transcripts were amplified in a parallel reaction using a specific primer pair (Wiede et al. 2001).

2.9.4
Antisera and Western Blot Analysis of TFFs

The following antisera-monitoring TFF peptides were used. Anti-TFF1: A polyclonal rabbit antiserum against the carboxy-terminal region of human TFF1 was purchased from Novocastra (Newcastle, UK) and used for Western blot analysis. Alternatively, a monoclonal antiserum against the 30 C-terminal amino acids of human TFF1 was purchased from Zymed Laboratories (San Francisco, Calif., USA) and used for immunohistochemistry. Anti-TFF2: The production of polyclonal rabbit antiserum anti-hTFF2-1 against the C-terminal region of TFF2 was described previously (Jagla et al. 2000). Anti-TFF3: The affinity-purified polyclonal rabbit antiserum anti-hTFF3-2 (Wiede et al. 2001) against the C terminus of human TFF3 was used for Western blot analysis, whereas a polyclonal rabbit antiserum against the C terminus of rat TFF3 (anti-rTFF3-1) (Wiede et al. 1999) was used for immunohistochemistry.

Human tissue (standardised ratio: 100 mg wet weight/400 μl buffer) was extracted under reducing conditions as described (Jagla et al. 1998) and the protein content measured by the Bio-Rad protein assay based on the Bradford dye-binding procedure. In total, 20 μg of protein was analysed by Western blot under reducing conditions as reported previously (Jagla et al. 1998), using the various antisera at a 1:1,000 dilution. Detection of the immunoreactivity was with the enhanced chemiluminescence (ECL) Western blotting analysis system (Amersham Pharmacia Biotech, England) using a Biomax MR film (Kodak, Rochester, N.Y., USA).

2.9.5
RT-PCR of Antimicrobial Peptides

For PCR, 4 μl cDNA were incubated with 30.5 μl water, 4 μl 25 mM $MgCl_2$, 1 μl dNTP, 5 μl 10×PCR buffer, and 0.5 μl (2.5 U) platinum *Taq* DNA polymerase (Gibco) and the following primers (2.5 μl each containing 10 pmol): BPI – bactericidal permeability-increasing protein (forward primer: 5' TTCCAGCTTCCCAGTTCCCAGATA 3'; reverse primer: 5' CATCCACGGCAGGGTAGAAGGTAA 3', 841 bp), CAP-37 – heparin binding protein (forward primer: 5' AGCATGAGCGAGAATGGCTACGAC 3'; reverse primer: 5' GGTCCTCGGGGGTCACAGTCA 3', 235 bp), LL-37 – human cationic antimicrobial protein (forward primer: 5' ATCATTGCCCAGGTCCTCAG 3'; reverse primer: 5' GTCCCCATACACCGCTTCAC 3', 251 bp), human alpha defensin 5 (HD5) (forward primer: 5' GCCATCCTTGCTGCCATTC 3'; reverse primer: 5' AGATTTCACACACCCCGGAGA 3', 241 bp), HD6 (forward primer: 5' CCTCACCATCCTCACTGCTGTTC 3'; reverse primer: 5' CCATGACAGTGCAGGTCCCATA 3', 269 bp), HBD-1 (forward primer: 5' TTGTCTGAGATGGCCTCAGGTGGTAAC 3'; reverse primer: 5' ATACTTCAAAAGCAATTTTCCTTTAT 3', 253 bp), HBD-2 (forward primer: 5' CCAGCCATCAGCCATGAGGGT 3'; reverse primer: 5' GGAGCCCTTTCTGAATCCGCA 3', 255 bp). Thirty-five cycles were done with each primer pair (annealing temperature 60°C), and 0.1 μM of a glyceraldehyde-3-phosphate dehydrogenase (GAPDH)-specific intron-spanning primer pair (5' CCAGCCGAGCCACATCGCTC 3'; 5' ATGAGCCCCAGCCTTCTCCAT 3'), which yielded a 360-bp amplified product, served as the internal control for equal amounts of cDNA. All primers were synthesised by MWG-Biotech AG (Ebersberg, Germany).

The positive control cDNA samples analysed included one sample from human blood (BPI, CAP 37, LL 37), one sample from cultured epithelial cells (HBD-1), and one sample from small intestine epithelium (HD5 and HD6). The cDNA was replaced with water for a negative control reaction.

2.10 Animal Investigations/Experiments

2.10.1 Animal Experiments with Rats

2.10.1.1 Iodination of Bovine Serum Albumin

In total, 60 μg (0.1 nmol) bovine serum albumin (BSA) was iodinated by the chloramine T method with 0.2 mCi (7.4 MBq) $Na^{125}I$ and purified from low-molecular-weight material by repeated (threefold) concentration and dilution with 0.14 M NaCl by ultrafiltration (YM 30 Microconcentrator from Amicon, Beverly, Mass., USA).

2.10.1.2 Absorption Experiments

One third of the iodinated BSA (approximately 50 μCi=1.85 MBq) in 20 μl 0.14 M NaCl each was applied by dropping into both eyes of female Wistar rats weighing 115–125 g. After 10, 20 and 60 min the rats were killed with carbon dioxide, decapitated, blood collected and the heads saved for histology.

2.10.1.3 Fractionation of Radioactivity in the Serum

Serum was obtained from the clotted blood (30 min) by centrifugation (12,000×*g*, 10 min) and aliquots analysed for protein-bound radioactivity by two methods. Serum (1 ml) was mixed (30 min) with 0.2 ml 3 M trichloroacetic acid (TCA), centrifuged (13,000×*g*, 10 min), the sediment washed with 1 ml 0.2 M TCA and the radioactivity in the sediment (protein-bound) and the combined supernatants (free radioactivity) counted. In a second approach, 1 ml serum was fractionated by gel chromatography and radioactivity in the fractions counted (Superdex S-200 superfine column 90×2.1 cm, Pharmacia, Uppsala, Schweden, equilibrated and eluted with 0.14 M NaCl 20 mM Hepes buffer, pH 7.4, at a flow rate of 0.3 ml/min, 2 ml fractions collected). In separate runs, the elution volumes of BSA and Tyr were determined by measuring their absorbance at 280 nm in the eluate.

2.10.1.4
Light Microscopy and Autoradiography

The heads were embedded in paraffin as described above and sectioned in a horizontal plane. Sections (7 μm) were stained with toluidine blue O (pH 8.5), alcian blue (pH 1.0), resorcin-fuchsine-thiazine-picric acid, and by the method of Goldner as well as mild periodic acid–Schiff (mPAS). For autoradiography, head sections (different horizontal planes) were placed on X-ray film (X-OMAT MA film 13×18 cm, Kodak, Paris, France) for 3 days, 1 week and 3 weeks.

2.10.2
Comparative Anatomy

2.10.2.1
Ocular Sample Preparation and Light Microscopy for Comparative Anatomy

Heads of apes, rabbits, rats, cats, deer and pigs were immediately fixed in 4% formalin no more than 24 h after death. Heads of rats were fixed for one week in 4% formalin, and than decalcified in 20% EDTA at 37°C for several days until x-ray examination revealed complete decalcification. Human nasolacrimal systems were removed from the bony channel prior to fixation and then also fixed in 4% formalin no more than 24 h after death. After 4 weeks of fixation, the nasolacrimal systems of apes, rabbits, cats, deer, and pigs were removed from the heads and post-fixed in 4% formalin. All human, primate, and vertebrate nasolacrimal systems were finally embedded in paraffin, and sectioned in a horizontal plane. Sections (7 μm) were stained with toluidine blue O (pH 8.5), alcian blue (pH 1.0), resorcin-fuchsine-thiazine picric acid, azan, and using the Goldner method.

2.10.2.2
Scanning Electron Microscopy for Comparative Anatomy

For scanning electron microscopy, four nasolacrimal ducts of rabbits and two of humans previously fixed in 4% formalin were cut longitudinally and halved to examine the lining epithelium. All tissue blocks were then impregnated with 2.5% tannic acid for 2 days. A counter-fixation in 2% OsO_4 for 4 h was followed by dehydration in ethanol and drying in a critical point dryer. The preparations were coated with gold and examined with a scanning electron microscope (Philips, Kassel, Germany).

2.10.3
Animal Experiments with Rabbits

2.10.3.1
Determination of ^{3}H-Cortisol Uptake

Of the cortisol, 250 μCi [1,2,6,7-^{3}H] (specific activity 100 Ci/mmol) was obtained from Amersham Pharmacia Biotech UK Limited, Buckinghamshire, UK (Code 117 B). The toluene-ethanol solution supplied was dried under a layer of nitrogen and the residue was dissolved in 80 μl 1% fatty acid-free bovine serum albumin in phosphate-buffered saline (0.14 M NaCl 20 mM Hepes buffer, pH 7.4). Of that, 20 μl (containing about 60 μCi ^{3}H-cortisol) were dropped into the eyes (10 μl into each eye) of female rabbits weighing 230–240 g using a Hamilton syringe. Then the eyelids were carefully closed by the investigator – 3/4 three times. After 21, 43 and 146 min the rabbits were killed with Narcoren (Merial GmbH, Hallbergmoos, Germany), decapitated, 10 ml blood was collected and the heads saved for autoradiography and histological examination. Serum was obtained from the clotted blood (30 min) by centrifugation (12,000×g; 10 min) and 2 ml aliquots were analysed for radioactivity by liquid scintillation using a quench standard.

2.10.3.2
Light Microscopy and Autoradiography

The nasolacrimal systems were removed from the heads of the rabbits. One system each was immediately fixed in 4% formalin, embedded in paraffin, and sectioned in a horizontal plane. Sections (7 μm) were stained with toluidine blue O (pH 8.5), alcian blue (pH 1.0), resorcin-fuchsine-thiazine picric acid, according to Goldner and with mPAS. The other nasolacrimal system in each case was divided into two parts, upper and lower. For autoradiography, these parts were placed on X-ray Hyperfilm ECL (Amersham Pharmacia Biotech, London, UK) for 3 weeks.

3 Results

3.1 The Epithelium and the Underlying Lamina Propria of the Nasolacrimal Ducts

3.1.1 Structure of the Epithelium

3.1.1.1 Histology

The internal wall of each lacrimal canaliculus was lined by a thick non-cornified stratified epithelium resting on a basement membrane. Subepithelial collagen fibrils and elastic fibres as well as muscle fibres of the lacrimal portion of the orbicularis oculi muscle were visible around the canaliculi (Fig. 2a).

At the transition from the canaliculus into the lacrimal sac, the stratified epithelium produced a double-layered epithelial area that lined the lacrimal sac and nasolacrimal duct and rested on a broad basement membrane (Fig. 2b, c). The two layers consisted of a basal-cell layer and a superficial columnar layer. In their apical part, epithelial cells of the columnar layer contained many vacuoles and lipid droplets (Fig. 2e, g). Goblet cells were integrated in the epithelium as solitary cells (Fig. 2b) or, particularly in the lacrimal sac, in a characteristic arrangement of several cell groups forming mucous glands (Fig. 2d, f). The secretory product of these cells reacted strongly positive with alcian blue (pH 1; Fig. 2d) and with mPAS. In areas with high goblet cell density, epithelial cells did not contain lipid droplets. Underneath the epithelium, the lamina propria consisted of two strata: (1) loose connective tissue containing a thin layer of elastic fibres and many lymphatic cells, sometimes arranged in follicles, as well as (2) a rich venous plexus situated under loose connective tissue that was connected caudally with the cavernous body of the nasal inferior turbinate. Besides epithelial and goblet cells, small seromucous glands (Fig. 2h) were situated in the region of the fornix of the lacrimal sac. Their excretory ducts penetrated the lining epithelium into the lumen of the sac.

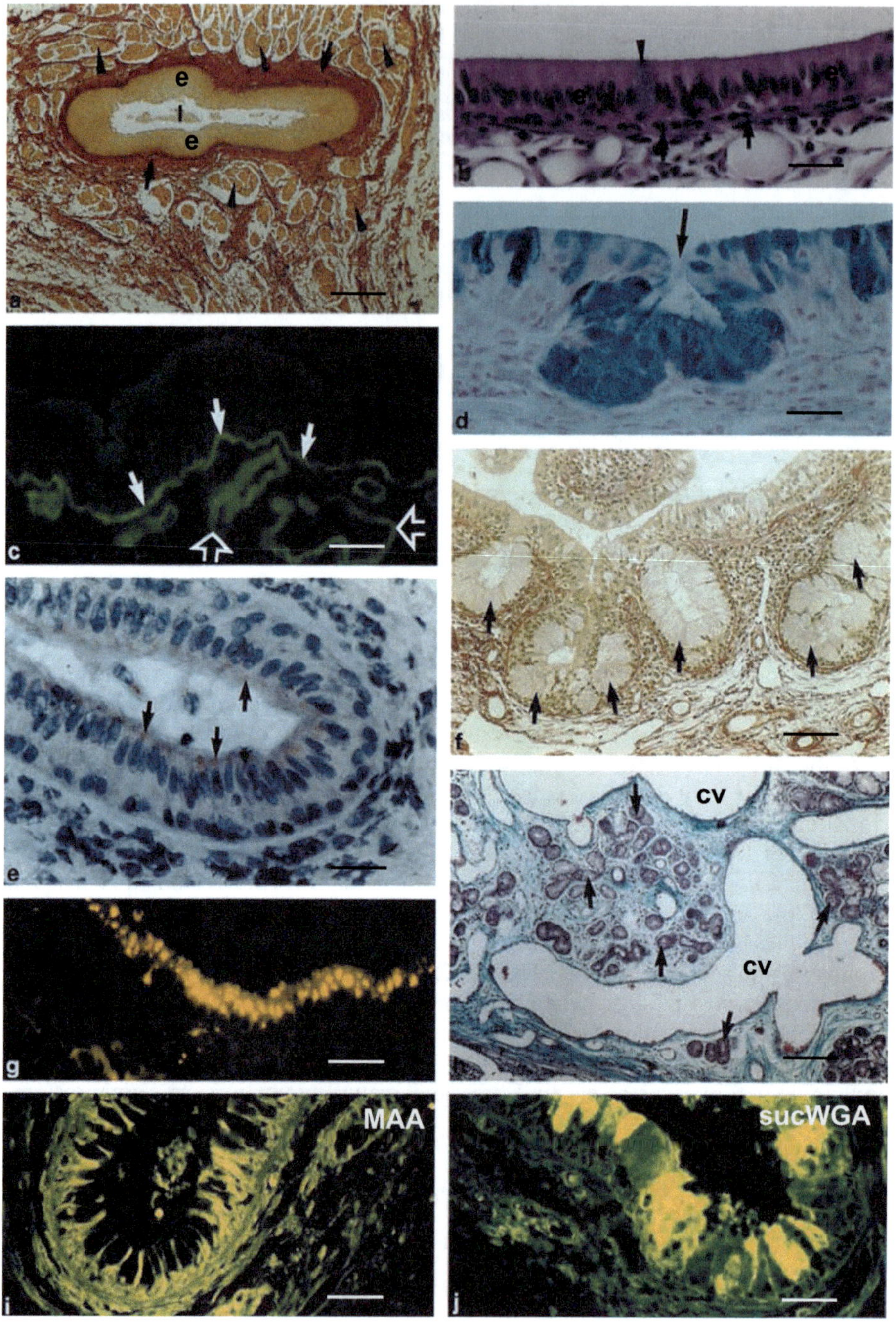
e
l
e
a
c
d
e
f
cv
cv
g
MAA
sucWGA
i
j

3.1.1.2
Ultrastructure of the Epithelium

Transmission electron microscopy (TEM) underlined the findings made by light microscopy. The lining of the lacrimal sac and the nasolacrimal duct revealed a double-layered epithelium consisting of a basal-cell layer and a superficial columnar layer (Fig. 3a). Besides a lot of mitochondria, cells of the superficial epithelial layer contained many vacuoles with a double-layered membrane and a lot of lipid droplets in their apical part (Fig. 3a). Cell borders between superficial epithelial cells were built of zonulae occludentes, zonulae adhaerentes and desmosomes. Golgi apparatus and endoplasmic reticulum could not be distinguished due to autolytic changes. Most of the epithelial cells were covered with microvilli (Fig. 3b). The length of the microvilli was between 400 and 600 nm, the diameter between 60 and 70 nm. An outer covering of a few tufts of kinocilia was seen in the lower part of the nasolacrimal duct.

Goblet cells were scattered among the epithelial cells and showed a columnar aspect. They were clearly distinguishable from epithelial cells on account of their differing cell surface. Microvilli-like cell processes were noticed on the cell surface (Fig. 3c). The nucleus was cup-like and located in the basal pole of the cells. Numerous mucous droplets filled the entire supranuclear cytoplasm and gave rise to the goblet-shaped appearance of the cell.

Below the basement membrane the lamina propria contained capillaries and many lymphocytes (Fig. 3b,g c).

Fig. 2a–j. Light microscopy of the lining epithelium of the nasolacrimal ducts. **a** Frontal section of a right lower canaliculus; (*l*) lumen, (*e*) epithelium, (*arrows*) collagen fibrils of lamina propria, (*arrowheads*) muscle fibres of the lacrimal portion of the orbicularis oculi muscle, resorcin-fuchsine-thiazine picric-acid staining. *Bar*=333 µm. **b** Epithelium (*e*) of the nasolacrimal duct consisting of a basal cell layer (*arrows*) and a superficial columnar layer. A goblet cell is integrated in the epithelium (*arrowhead*). *Bar*=42 µm. **c** Immunohistochemical proof of laminin. Reactivity is visible along the basement membrane (*arrows*). Additionally the antibody marks subepithelial blood vessels (*arrows*) of the nasolacrimal ducts. *Bar*=27 µm. **d** Horizontal section through the lining epithelium of the lacrimal sac. Goblet cells form a mucous gland that lies in the lamina propria and has its own secretory duct (*arrow*). Alcian blue staining (pH 1); *bar*=27 µm. **e** Horizontal section though the lacrimal sac. Intracellular lipids stain yellow-red with Sudan-III. The lipids are distributed symmetrically intracellular above the nucleus (*arrows*) of epithelial cells. Counterstaining with haemalum; *bar*=27 µm. **f** Goblet cells show a characteristic arrangement of several cell groups in the upper part of the lacrimal sac forming mucous glands (*arrows*). Resorcin-fuchsine-thiazine picric-acid staining; *bar*=84 µm. **g** Immunohistochemical proof of chondroitine-6-sulphate. Using a filter of 450- to 490-nm wavelength, intraepithelial lipids show a yellow fluorescence. *Bar*=27 µm. **h** Horizontal section through the vascular system of the lacrimal sac. The lamina propria contains seromucous glands (*arrows*); *cv*, lumen of capacitance veins of the vascular system; Goldner staining; *bar*=84 µm. **i** Histochemical staining with the lectin *Maackia amurensis* agglutinin (*MAA*). MAA staining is restricted to epithelial cells and does not stain goblet cells. *Bar*=27 µm. **j** Histochemical staining with the lectin succinylated wheat germ agglutinin (*sucWGA*). SucWGA binding is uniformly distributed throughout goblet cells, whereas epithelial cells are completely sucWGA-negative. *Bar*=27 µm

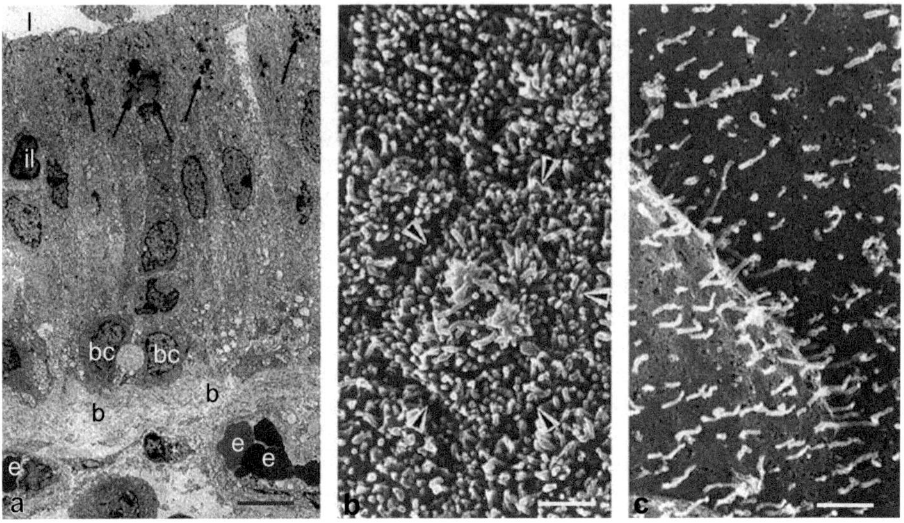

Fig. 3a–c. Transmission electron microscopy (TEM) and scanning electron microscopy (SEM) of the lining epithelium of the lacrimal sac and the nasolacrimal duct. **a** TEM micrograph showing a section through the epithelium of the lacrimal sac. The preparation is not taken until 48 h of death and shows autolytic changes. *Arrowheads,* subepithelial blood vessels; *e,* erythrocytes; *b,* basement membrane; *bc,* basal cells; *il,* intraepithelial lymphocyte; *arrows,* intraepithelial lipid droplets; *l,* lumen of the lacrimal sac; *bar*=8.7 µm. **b** SEM micrograph of the surface of epithelial cells in the lacrimal sac. Cell borders are clearly distinguishable (*arrowheads*). SEM reveals a surface covering of epithelial cells by a trimming with microvilli. *Bar*=2 µm. **c** SEM micrograph of the surface of goblet cells in the lacrimal sac. Cell borders are clearly distinguishable. SEM reveals a surface covering of the goblet cells by a very sparse trimming with microvilli-like cell processes. *Bar*=5 µm

3.1.2 Receptor Binding of Carbohydrate-Specific Lectins to the Epithelium

Application of carbohydrate-specific lectins (ConA, GSA-I, GSA-II, LPA, MAA, SNA, UEA-I, WGA, sucWGA, MPA) on deparaffinated tissue sections of healthy lacrimal duct epithelium of the lacrimal sac and the nasolacrimal duct resulted in different staining patterns of goblet cells and epithelial cells inside the epithelium (Table 2). Binding sites for UEA-I-, WGA-, sucWGA- and MPA-binding were uniformly distributed throughout goblet cells (Fig. 2i), whereas epithelial cells were UEA-I-, WGA-, sucWGA and MPA-negative (Fig. 2j; shown for sucWGA). In contrast, MAA staining was restricted to epithelial cells and did not show binding sites in goblet cells (Fig. 2i). SNA binding was detected in both goblet cells and epithelial cells. No specific staining was obtained with GSA-I, GSA-II or for ConA. No staining was seen with LPA.

Conclusions (3.1.1–3.1.2). Synthesised mucins of goblet cells and epithelial cells form a specialised protective layer on the epithelium of the nasolacrimal ducts, which functionally could serve for a simplified drainage of tear fluid into the inferior meatus of the nose. Together with immunocompetent cells, the protective layer plays

Table 2. Summary of lectin binding in efferent tear duct epithelium

Lectin	Binding to goblet cells	Binding to epithelial cells
Concanavalin A agglutinin	+ (non-specific)	+ (non-specific)
Griffonia simplicifolia I agglutinin	–	+ (non-specific)
Griffonia simplicifolia II agglutinin	–	+ (non-specific)
Limulus polyphemus agglutinin	–	–
Maackia amurensis agglutinin	–	+
Sambucus nigra agglutinin	+	+
Ulex europaeus I agglutinin	+	–
Wheat germ agglutinin	+	–
Succinylated wheat germ agglutinin	+	–
Maclura pomifera agglutinin	+	–

a role in antigen defence, prevents invasion of pathogenic agents and obviates dacryocystitis as well as the formation of dacryostenosis. The facing of epithelial cells by microvilli gives hints of reabsorption of lacrimal fluid components inside the nasolacrimal ducts. Moreover, demonstration of intraepithelial lipids and mucins speak for maintenance of the normal three-layering of the tear film inside the nasolacrimal passage.

3.1.3 Trefoil Factor Peptides in the Epithelium

3.1.3.1 RT-PCR Analysis

RNA was isolated from surgically obtained nasolacrimal duct tissue of 18 patients and from the nasolacrimal ducts of four body donors of which cDNA was amplified by the use of specific primer pairs (Wiede et al. 2001), testing for TFF1, TFF2 or TFF3 transcripts (Fig. 4). As a control, GAPDH transcripts were amplified. TFF1-specific amplification products were visible at a low level after separation on an agarose gel in nearly half of the investigated specimens, whereas TFF3-specific amplification products were clearly detected in every sample. By contrast, expression of TFF2 was not detectable in human nasolacrimal ducts. As controls, TFF1, TFF2 and TFF3 transcripts were monitored in parallel reactions with cDNA samples from human stomach and colon (Fig. 4).

3.1.3.2 Western Blot Analysis

Nasolacrimal ducts or lacrimal sacs were dissected from different individuals and extracts were tested for TFF peptides using Western blot analysis (Fig. 5). Relatively small amounts of TFF1 could be detected in all samples. In contrast, the TFF3 content differed greatly between the samples. There is a tendency towards higher TFF3

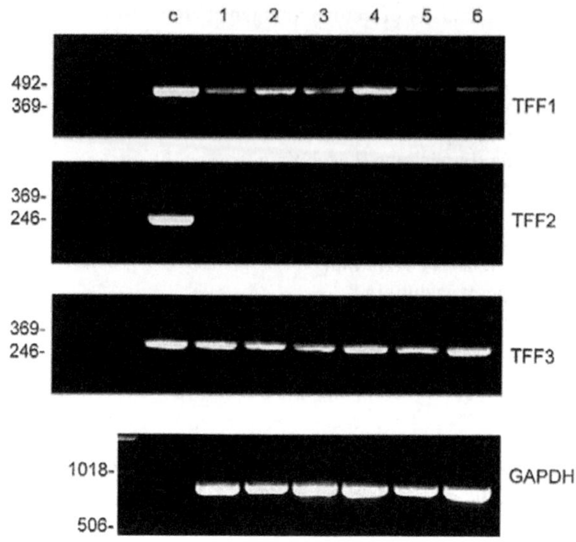

Fig. 4. RT-PCR analysis TFF1, TFF2 or TFF3 expression was monitored in the human lacrimal epithelium of 6 individuals (2 body donors and 4 specimens from surgical material). Total RNA from stomach or colon was analysed as positive controls for TFF1, TFF2 or TFF3 transcripts (*lane c*). The integrity of the cDNAs was tested by amplification of the GAPDH transcript. The molecular size standard is shown at the *left*

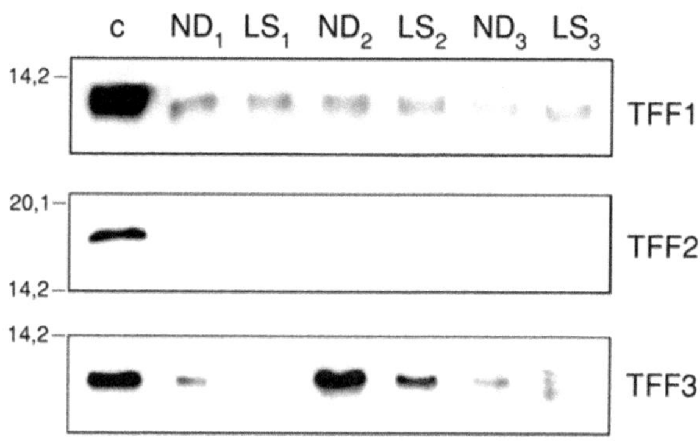

Fig. 5. Detection of TFF peptides in post-mortem material of nasolacrimal ducts (*ND*) or lacrimal sacs (*LS*). Extracts (20 µg total protein) from three different individuals were analysed by SDS-polyacrylamide gel electrophoresis (15%) and subsequent Western blot using the following polyclonal antisera against TFF peptides: anti-human TFF1 (*TFF1*), affinity-purified anti-hTFF2-1 (*TFF2*), or affinity-purified anti-hTFF3-2 (*TFF3*). Human stomach or duodenum extracts were used as positive controls (*c*). The molecular size standard is shown at the *left*

concentrations in the nasolacrimal ducts as compared with the lacrimal sacs from the same individuals. TFF2 was not detectable at all using this assay.

3.1.3.3 General Histology and Immunohistochemistry

Reactivity of TFF1 was observed in nearly half of the specimens investigated; TFF3 was detectable in each specimen. The cellular localisation of TFF1 and TFF3 revealed a completely different pattern. If present, staining of TFF1 was found mostly only perinuclearly in goblet cells organised as intraepithelial glands and sometimes also weakly in their cytoplasm (Fig. 6a). Although TFF1 staining looked like nuclear staining in several cases (compare Fig. 6d), scrolling with the micrometer-screw of the microscope through the sections showed clearly that the staining was perinuclear and not nuclear. Moreover, the staining was highly specific and was inhibited with the corresponding synthetic peptide (Fig. 6b). The TFF1-positive cells were also alcian blue-positive due to their characteristic mucin contents (Fig. 6c). TFF3 occurred in epithelial cells of the lacrimal sac and the nasolacrimal duct revealing strong staining (Fig. 6e, f) as well as in acinar cells of subepithelial serous glands (Fig. 6g). Reactivity for TFF3 was also visible perinuclearly in goblet cells but was absent in their stored secretory product (Fig. 6f). Moreover, TFF3 was visible in cells inside the lamina propria which had a fibroblast-like appearance. There were variations in intensity between different epithelial cells, goblet cells and especially serous cells of subepithelial serous glands. The latter revealed strong nuclear staining of some serous cells in single sections (Fig. 6g). Competition experiments with the corresponding synthetic peptide revealed that the nuclear staining could not be inhibited, whereas the cytoplasmic staining of TFF3 was inhibitable (Fig. 6h). Double staining of TFF1 and TFF3 revealed that both peptides were related to different cell types (Fig. 6d): TFF1 to goblet cells forming intraepithelial mucous glands and TFF3 to columnar epithelial cells. TFF2 was not detected in all the samples of the nasolacrimal ducts tested using the polyclonal antiserum anti-hTFF2-1 that successfully detected TFF2 in human stomach biopsies (data not shown).

Conclusions (3.1.3). The epithelium of the nasolacrimal ducts synthesises TFF3 and in some cases also TFF1. In contrast to the human conjunctiva, where TFF3 is only

Table 3. Detection of antimicrobial peptides by RT-PCR in surgically obtained biopsies of nasolacrimal ducts of patients suffering from different types of tear duct stenosis with and without dacryocystitis

No.	Sex	Age	Ad	BPI	CAP37	LL37	HD5	HD6	HBD1	HBD2
1	F	66	–	+	+	–	–	–	+	–
2	F	78	–	+	+	–	–	–	+	–
3	M	5	–	+	+	–	–	–	+	–
4	F	6	–	+	+	–	–	–	+	–
5	F	44	aDC	+	–	–	–	–	+	+
6	F	51	–	+	+	–	–	–	+	–
7	F	58	aDC	+	+	–	–	–	+	+
8	F	44		+	+	–	–	–	+	–
9	M	64	sDC	+	+	–	–	–	–	+
10	M	66	–	+	+	–	–	–	+	–
11	F	73	cDC	+	+	–	–	–	–	+
12	M	63	–	+	+	–	–	–	+	–
13	F	66	aDC	+	–	–	–	–	–	+
14	F	84	aDC	+	–	–	–	–	–	+
15	F	57	–	+	+	–	–	–	+	–

Ad, additional inflammation; aDC, acute dacryocystitis; BPI, permeability-increasing protein; CAP37, heparin-binding protein; cDC, chronic dacryocystitis; HBD1, human β-defensin 1; HBD2, human β-defensin 2; HD5, human α-defensin 5; HD6, human α-defensin 6; LL37, human cationic antimicrobial protein; sDC, subacute dacryocystitis.

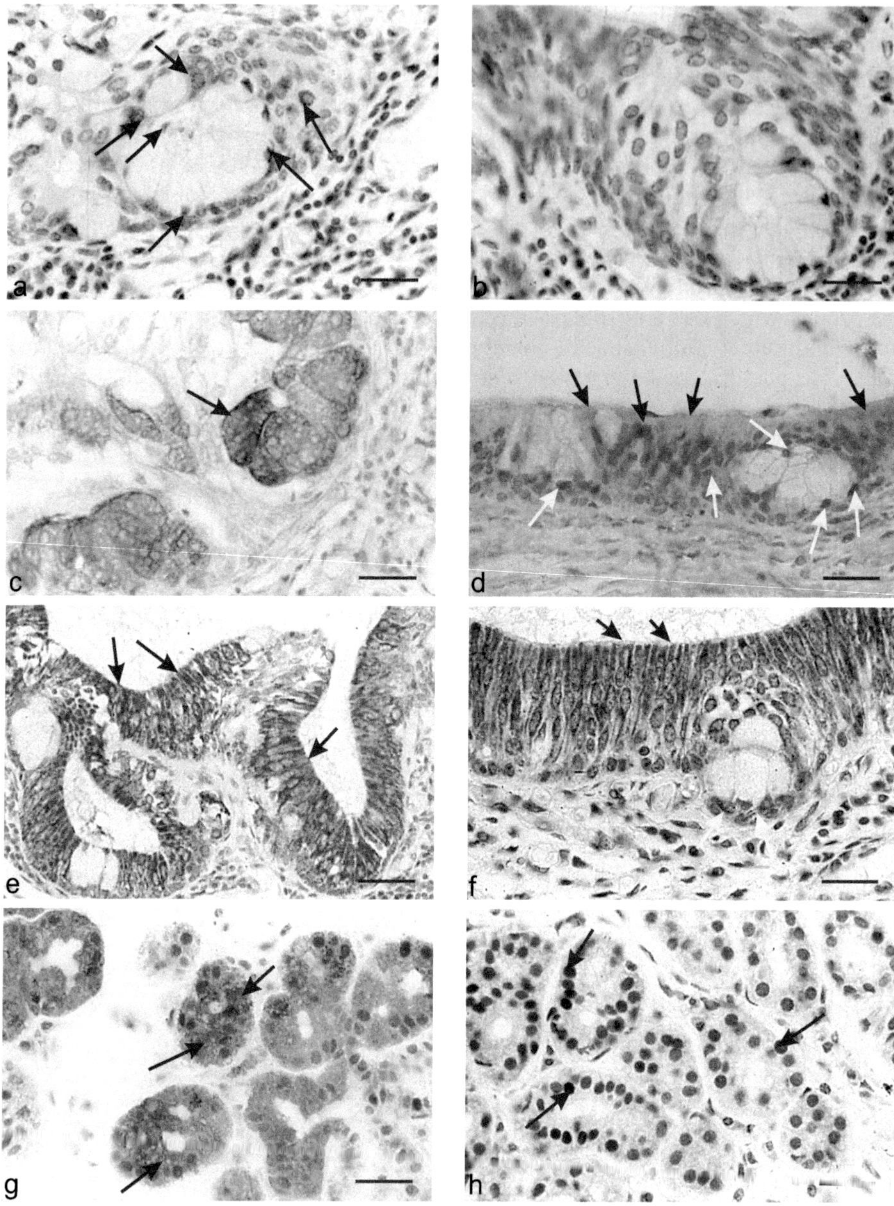
a
b
c
d
e
f
g
h

detectable in goblet cells, TFF3 of the nasolacrimal ducts is produced in large amounts by epithelial cells as well as by serous glands, but not – or in small amounts only – by goblet cells. This is comparable with TFF3 localisation in the major salivary glands. Thus, TFF3 may have a special function in tear transport through the lacrimal passage comparable to its function on the ocular surface because the peptide, together with TFF1, may contribute to the rheological properties of the tear film. Moreover, the TFF peptides may also influence epithelial healing due to their motogenic properties.

3.1.4 Localisation of Antimicrobial Peptides

3.1.4.1 RT-PCR Analysis

RT-PCR of normal non-inflamed and inflamed nasolacrimal duct samples for BPI revealed a product at 841 bp in all samples analysed (Tables 3 and 4; Fig. 7a). The negative control reaction (DNA-free water instead of cDNA) showed no PCR products. Internal control (GAPDH) revealed a product at 360 bp.

Products of 253 bp (CAP37) and 253 bp (HBD-1) revealed both peptides in normal non-inflamed and inflamed nasolacrimal duct samples (Fig. 7a, b; Tables 3 and 4). Positive control cultured epithelial cells also revealed corresponding products of 235 bp and 253 bp (Fig. 7b). The negative control reactions (DNA-free water instead of cDNA) showed no PCR products. The internal control revealed the expected 360-bp product.

Three mucosal samples revealed no expression of CAP37 (biopsy samples 5, 13, and 14 from Table 3). Four mucosal samples (9, 11, 13, and 14; Table 3) showed no HBD-1 expression. It was notable that all the patients in these cases had acute or chronic dacryocystitis, and all the mucosal samples revealed expression of HBD-2 (Table 3).

Fig. 6a–h. TFF1 and TFF3 in human nasolacrimal ducts. **a** Localisation of TFF1 (*arrows*) to goblet cells of the lacrimal sac using the monoclonal TFF1 antiserum. *Bar*=27.5 µm. **b** Competition with the corresponding synthetic peptide of TFF1 reveals that the staining can be inhibited (compare with Fig. 3a) *Bar*=27.5 µm. **c** Parallel section to part *a* stained with alcian blue (pH 2.5) which stains goblet cells (*arrow*); counterstaining was with haemalum. **d** Double staining of TFF1 (*white arrows*) and TFF3 (*black arrows*) in human nasolacrimal ducts. No co-localisation of TFF1 and TFF3 is visible. TFF1 is related to goblet cells whereas TFF3 stains columnar epithelial cells. *Bar*=27.5 µm. **e**, **f** Localisation of TFF3 (*arrows*) to columnar epithelial cells of the lacrimal sac (*section e*) (*bar*=43.3 µm) and the nasolacrimal duct (*section f*) (*bar*=27.5 µm) using antiserum anti-rTFF3-1. **g** Localisation of TFF3 to serous cells of serous glands using antiserum anti-rTFF3-1 (*arrows*). *Bar*=27.5 µm. **h** Competition with the corresponding synthetic peptide of TFF3 reveals that the nuclear staining (*arrows*) visible in *g* cannot be inhibited, whereas the cytoplasmic staining of TFF3 is inhibitable. *Bar*=27.5 µm

Fig. 7a–c. mRNA expression of antimicrobial peptides in nasolacrimal ducts. RT-PCR analyses performed with RNA samples obtained from different tissues. **a** Detection of BPI and CAP37 mRNA in the lacrimal epithelium of a body donor (*lane 1*) and a patient (*lane 2*). **b** Detection of HBD-1 message in a Hakard epithelial cell line (which served as control; *lane 1*), in the lacrimal epithelium of a body donor (*lane 2*) and a patient (*lane 3*). **c** Detection of HBD-2 mRNA in the lacrimal epithelium of a patient (*lane 1*) who had suffered from subacute dacryocystitis. GAPDH

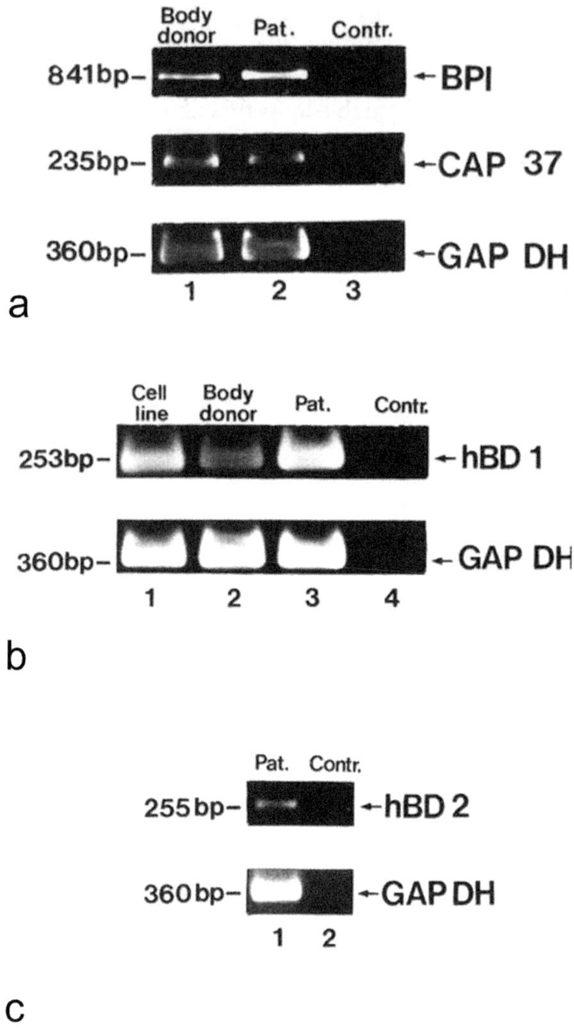

RT-PCR of all cadaveric samples was negative for HBD-2 (Table 4), with one exception, revealing a weak HBD-2 product of 255 bp. In contrast, all the inflamed nasolacrimal duct samples obtained by surgery revealed a product at 255-bp product for HBD-2 (Fig. 7c, Table 3). A positive control was not available for HBD-2. The negative control showed no PCR products; the internal controls revealed the expected 360-bp product for GAPDH.

RT-PCR of all non-inflamed and inflamed samples was negative for LL 37, as well as for HD5 and HD6. Human blood and small intestine samples showed appropriate PCR products (Tables 3 and 4).

Table 4. Detection of antimicrobial peptides by RT-PCR in biopsies of nasolacrimal ducts from body donors

No.	Sex	Age	BPI	CAP37	LL37	HD5	HD6	HBD1	HBD2
1	M	70	+	+	–	–	–	+	–
2	M	78	+	+	–	–	–	+	–
3	M	53	+	+	–	–	–	+	–
4	F	88	+	+	–	–	–	+	–
5	F	64	+	+	–	–	–	+	–
6	F	51	+	+	–	–	–	+	+
7	F	58	+	+	–	–	–	+	–
8	F	67	+	+	–	–	–	+	–
9	M	64	+	+	–	–	–	+	–
10	M	66	+	+	–	–	–	+	–

BPI, permeability-increasing protein; CAP37, heparin-binding protein; HBD1, human β-defensin 1; HBD2, human β-defensin 2; HD5, human α-defensin 5; HD6, human α-defensin 6; LL37, human cationic antimicrobial protein.

3.1.4.2 Antimicrobial Peptide Immunostaining

Lysozyme and lactoferrin were present in both the epithelium of the lacrimal sac and nasolacrimal duct and in the seromucous glands of the lacrimal sac. Lysozyme was produced by some of the epithelial cells, rendered visible as a red staining of the entire cytoplasm (Fig. 8a). Lactoferrin occurred in some of the epithelial cells as a fine granulation throughout the upper part of the cell (Fig. 8c). In seromucous glands, both antibodies stained the serous cells (Fig. 8b). Secretory phospholipase A_2 occurred only in epithelial cells of the epithelial lining, visible as a red staining of small intraepithelial vesicles (Fig. 8d). Production of lysozyme, lactoferrin and phospholipase A_2 was detected in normal uninflamed epithelium as well as in samples of inflamed epithelium.

Immunohistochemistry of the inflamed nasolacrimal duct samples with Def 3 showed strong positive staining of masses of neutrophils in the substantia propria and epithelium. However, all of the normal uninflamed nasolacrimal duct samples also revealed red staining of scattered subepithelial and intraepithelial neutrophils (Fig. 8e). Moreover, epithelial cells of the lacrimal passage also showed a weak staining of the entire cytoplasm. The seromucous glands revealed staining of the ductular epithelial cells with Def 3 and high-power views demonstrated staining of intracellular granules within the lining epithelial cells of the ductules.

Immunoreactivity of HBD-1 was visible only in some of the healthy and inflamed samples. When present, it occurred as a fine granulation throughout the upper part of the epithelial cells (Fig. 8f). Immunoreactive HBD-2 peptide was only visible in some of the inflamed samples (Fig. 8g), rendered visible as a red staining of the entire cytoplasm of the epithelial cells (Fig. 8h).

Conclusions (3.1.4). The data suggest that the human efferent tear ducts produce a broad spectrum of antimicrobial peptides. Under inflammatory conditions, changes in the expression pattern occur, revealing induction of the human inducible defensin

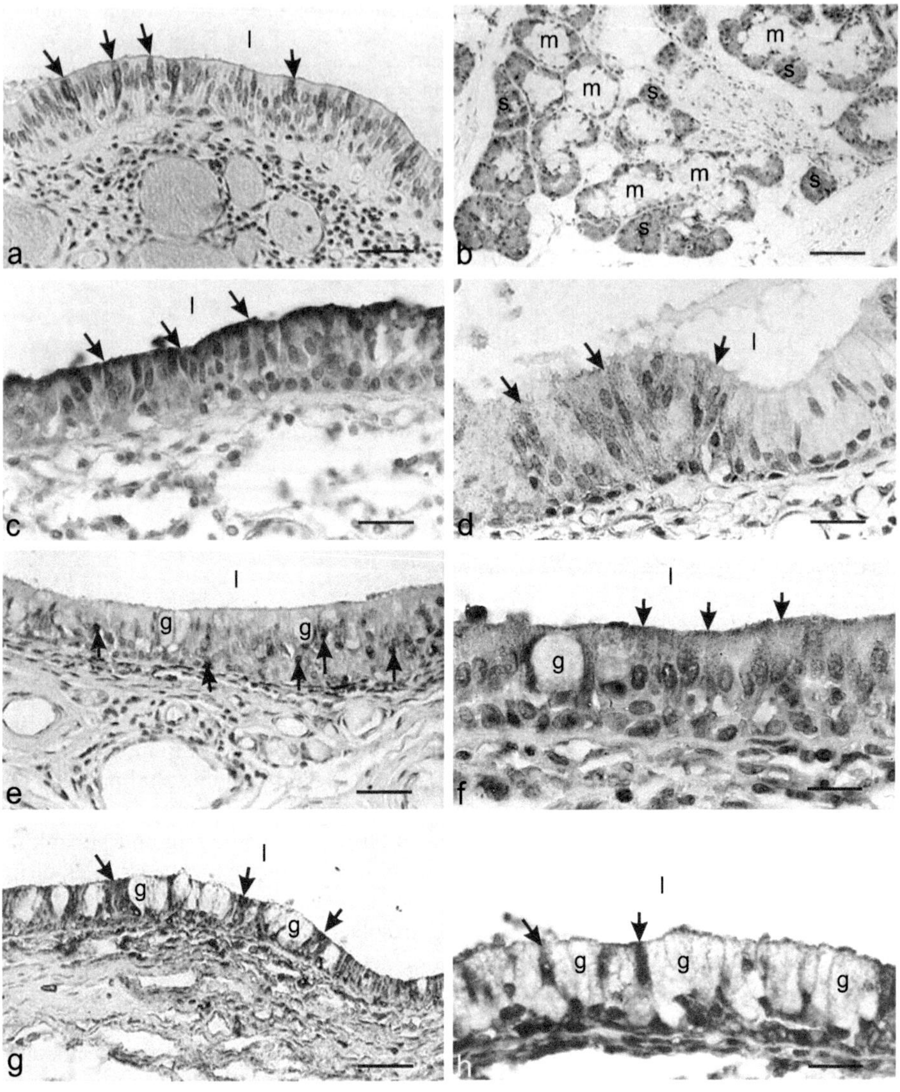
l
a
m
s
b
l
c
l
d
l
g
e
l
g
f
l
g
g
l
g
h

HBD-2 and in some cases downregulation of HBD-1 and CAP37. Antimicrobial peptides hold therapeutic potential in dacryocystitis as they have a broad spectrum of antimicrobial activity and accelerate epithelial healing. However, caution is appropriate since defensins also promote fibrin formation and cell proliferation, which are key elements in scarring processes, such as dacryostenosis.

3.1.5 Mucosa Associated Lymphoid Tissue

3.1.5.1 Distribution and Organisation in "Healthy" (Cadavers) Nasolacrimal Systems

To ensure comparability with surgically obtained specimens, only those parts of lacrimal systems from cadavers that corresponded to those obtained surgically were cut serially and investigated. The possibility that cases classified as "without MALT" did contain MALT in other parts of the lacrimal system can therefore not be excluded for both samples from cadavers and surgically obtained specimens.

The samples from cadavers exhibited a pseudo-stratified columnar epithelial lining consisting of a basal cell layer and a superficial columnar layer in most areas (Fig. 9a). Goblet cells were integrated in the epithelium as solitary cells or as intraepithelial mucous glands (Fig. 9a). Beneath the epithelium, the lamina propria consisted of loose connective tissue containing many lymphocytes (Fig. 9a), sometimes arranged in follicles and an underlying cavernous body. Small subepithelially located seromucous glands were detected in some samples in addition to epithelial and goblet cells (Fig. 9b).

Fig. 8a–h. Immunohistochemistry of antimicrobial peptides. **a** Horizontal section through the lining epithelium of the nasolacrimal duct. Some epithelial cells (*arrows*) react positively with an antibody against lysozyme. *Bar*=87 μm. **b** Immunohistochemical proof of lysozyme protein expression in seromucous glands of the lacrimal sac. Lysozyme staining is restricted to serous parts (*s*) of seromucous glands. Mucous parts (*m*) show no reactivity. *Bar*=161 μm. **c** Horizontal section through the lining epithelium of the nasolacrimal duct. The apical parts of most epithelial cells (*arrows*) react positively with an antibody against lactoferrin. *Bar*=27.5 μm. **d** Horizontal section through the lining epithelium of the lacrimal sac. Small intraepithelial vesicles of epithelial cells (*arrows*) react positively with an antibody against secretory phospholipase A_2. *Bar*=27.5 μm. **e** Horizontal section through the lining epithelium of the nasolacrimal duct of a body donor. Intraepithelial neutrophils (*arrows*) react positively with an antibody against α-defensins 1–3. *Bar*=87 μm. **f** Horizontal section through the lining epithelium of the nasolacrimal duct of a body donor. The apical parts of most epithelial cells (*arrows*) react positively with an antibody against HBD-1. *Bar*=27.5 μm. **g** Horizontal section through the lining epithelium of the lacrimal sac of a patient who suffered from dacryocystitis. Epithelial cells reveal strong red staining of the entire cytoplasm (*arrows*) with an antibody against HBD-2. *Bar*=87 μm. **h** Horizontal section through the lining epithelium of the nasolacrimal duct of a patient who suffered from dacryocystitis. Positively reacting epithelial cells reveal strong staining of the entire cytoplasm (*arrows*) with an antibody against HBD-2. *Bar*=27.5 μm; *l*, (in *sections a, c–h*) lumen of the lacrimal passage; *g*, (in *sections e–h*) intraepithelial goblet cells

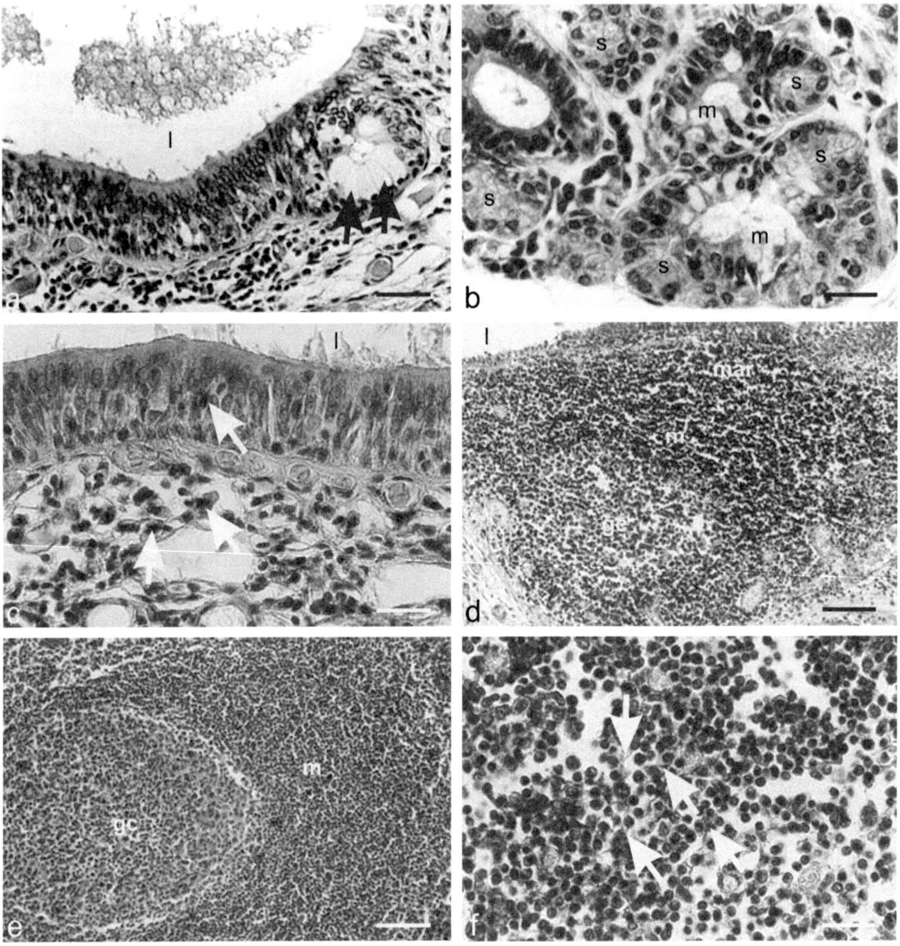

Fig. 9a–f. Histology and immunohistochemistry of lymphoid tissue. **a** Cross-section through the lining epithelium of the lacrimal sac. The goblet cells show a characteristic arrangement with several cells forming an intraepithelial mucous gland (*arrows*). Subepithelially, the lamina propria consists of loose connective tissue containing many lymphocytes. *l*, Lacrimal passage, haemalum staining; *bar*=52 µm. **b** Horizontal section through a small gland in the lamina propria of the lacrimal sac. The gland consists of serous (*s*) and mucous (*m*) parts. Haemalum staining; *bar*=27.5 µm. **c** Immunostaining with KiB3 of a horizontal section through the lining epithelium of the lacrimal sac of a cadaver (male, 88 years). Red staining reveals intraepithelial and subepithelial CD45RA-positive B cells (*arrows*). Immunostain – anti-CD45RA; *bar*=38 µm. **d** Immunostaining with CD20. Tear duct-associated lymphoid tissue (TALT) in a patient (female, 68 years) at the area of the prestenotic rim. TALT of the lacrimal sac with a well-developed germinal centre (*gc*), a mantle zone (*m*), and a marginal zone (*mar*). Most of the cells stain positive with an antibody against CD20. *l*, Lacrimal passage; azan staining; *bar*=113 µm. **e** Section through a well-developed germinal centre (*gc*) and a mantle zone (*m*) of tear duct-associated lymphoid tissue in a patient (female, 48 years) at the area of the poststenotic rim. Azan staining; *bar*=113 µm. **f** Anti-KiM4 immunostaining shows red staining of dendritic cells (*arrows*) in a germinal centre within a lymphoid follicle of the nasolacrimal duct of a patient (male, 32 years). Immunostain – anti-follicular dendritic cells. *Bar*=38 µm

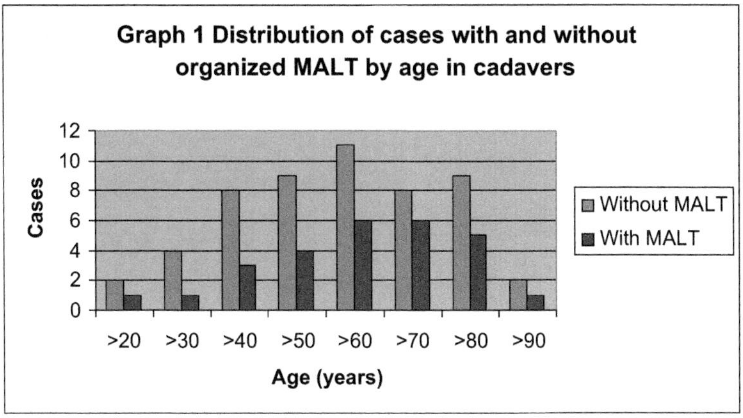

Fig. 10. Distribution of cases with and without organised MALT by age in cadavers

In most of the cases investigated there was a diffuse infiltrate of variable intensity within the lamina propria of the samples consisting predominantly of CD3-positive T lymphocytes with scattered CD20- and CD45RA-positive B cells as well as plasma cells. A few IgM- and IgG-positive plasma cells were also detected in the infiltrate and in the connective tissue between seromucous glands. Solitary CD3-positive T cells and CD45RA-positive B cells were detected inside the epithelium (Fig. 9c), although the overall extent of infiltration was quite low.

Organised lymphoid tissue (Fig. 9a) was identified in about one-third of the cases investigated. The findings are summarised in Fig. 10. The youngest person in whom MALT was identified was 27 years old.

MALT was found once in a right lacrimal sac and 4 times in a left one, as well as 13 times in a right nasolacrimal duct and 9 times in a left one. Either it was distributed at intervals along the investigated segment or only a single follicle was found. MALT was not necessarily present to the same extent in both efferent lacrimal systems.

3.1.5.2 Distribution and Organisation in "Inflamed" (Surgically Obtained) Nasolacrimal Systems

All 38 specimens analysed by histologic examination revealed chronic inflammation of the lacrimal system which had led to different stages of fibrosis, ranging from proliferative sclerotic forms of chronic fibrosis ("moderate" – 18 cases) to total subepithelial fibrosis ("severe" – 20 cases).

Tissue specimens from 18 patients demonstrated transitional forms of chronic fibrosis with epithelial and subepithelial infiltration by defence cells that were almost lymphocytes, loss of intraepithelial goblet cells, variable loss of differentiated epithelial cells ranging from denuded epithelium to basal cell hyperplasia – often associated with squamous metaplasia – and increased subepithelial fibroblasts (see Fig. 26a,

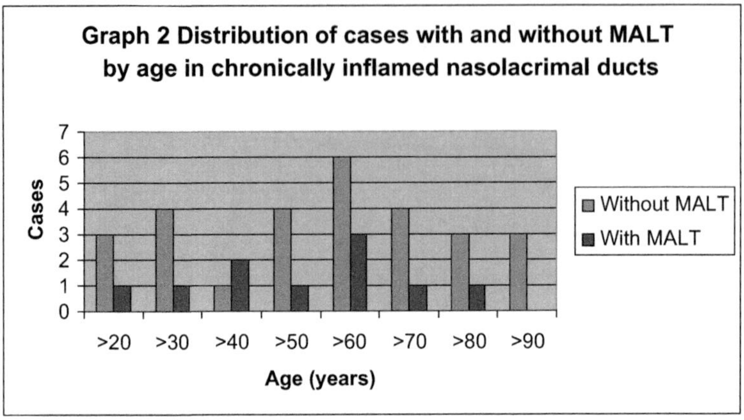

Fig. 11. Distribution of cases with and without MALT organised by age in chronically inflamed nasolacrimal ducts

b). The specialised blood vessels of the subepithelial cavernous body often showed thickening of the walls with varying degrees of intimal proliferation up to obliteration and reduction (see Fig. 26c, d).

Seven cases were characterised by squamous metaplasia with a total absence of goblet cells. The underlying mucosa had undergone secondary fibrosis with basement membrane thickening (see Fig. 26b). Also visible was a reduction – up to total loss – of the specialised blood vessels of the cavernous body (see Fig. 26c, d). Tissue specimens from 13 cases showed complete fibrous obliteration of the duct within the osseous canal (no lacrimal passage lumen was found, either intraoperatively or histologically).

The dissection rims of all surgically obtained specimens showed a transition from the described changes to a normal double-layered epithelium with integrated goblet cells and intraepithelial mucous glands. The lamina propria contained many lymphocytes, in seven cases arranged in follicles, and an underlying cavernous body.

Comparable to the cadavers, there was a diffuse infiltrate of variable intensity within the lamina propria of the surgically obtained specimens. The lamina propria consisted of loose connective tissue containing many lymphocytes, sometimes arranged in follicles and an underlying cavernous body. However, the diffuse infiltrate was only observed in the 18 specimens classified as "moderate." The specimens classified as "severe" revealed such a diffuse infiltrate only at their dissection rims. The central part consisted of fibrous tissue and was nearly free of detectable defence cells. Only single macrophages or lymphocytes were visible.

Organised lymphoid tissue was identified in 10 of the cases investigated. Of these, 4 belonged to the "moderate" group and 6 to the "severe" group. The findings are summarised in Fig. 11. The youngest person in whom MALT was identified is a 22-year-old. In 4 cases (2 "moderate", 2 "severe") MALT was situated in the area of the prestenotic rim; in 6 cases (2 "moderate", 4 "severe") it was detectable at the poststenotic rim.

3.1.5.3
Organisation of MALT in Cadavers and Surgically Obtained Specimens

It must be stated that only secondary follicles revealing a distinct germinal centre and a surrounding parafollicular area were counted (Figs. 9d, e, 12a, b). Several areas were seen to display more or less developed primary follicles. Such areas were excluded and named MALT by definition.

No significant male–female differences were seen to obtain between lacrimal systems from cadavers and from surgically obtained specimens. Comparable to other mucosal sites, MALT in the lacrimal passage of both cadavers and surgically obtained specimens was characterised by the presence of reactive germinal centres containing tingible body macrophages (Fig. 12b), a network of KiM4-positive follicular dendritic cells (Fig. 9f) and CD3-positive T cells (Fig. 12d). The germinal centres were surrounded by mantle zones and marginal zone cells (Figs. 9d, e, 12a, b). The mantle zones consisted of small CD20-positive lymphocytes (Fig. 9d) expressing KiB3 and IgM. These lymphocytes merged in a population of small-to-medium-sized B cells with moderately abundant cytoplasm and irregular nucleus outlines, features typical of marginal zone cells (Figs. 9d, 12b). The marginal zone cells exhibited the following immunophenotypes: $CD20^{+}$, $CD45RA^{+}$, $CD3^{-}$, IgM^{+} and IgA^{+}. Some IgG-expressing cells were present, although their occurrence was rare. The marginal zone cells extended into the overlying epithelium to form a characteristic lymphoepithelium (Fig. 12a, b, e). In the parafollicular area, mainly T lymphocytes, some B lymphocytes and macrophages were present as well as high endothelial venules. In no case were features suggestive of lymphoma observed.

3.1.5.4
IgA Immunostaining in "Healthy" (Cadavers) Nasolacrimal Systems

Strong immunoreactivity for IgA was observed in the plasma cells of the lamina propria beneath the epithelium (Fig. 12c), in the plasma cells of the connective tissue between seromucous glands and in the secretory products of the mucous membrane and the seromucous glands forming a layer on the lining epithelium. A somewhat weaker immunoreactivity for IgA was seen in the apical epithelial cells (Fig. 12c) and the serous cells of the seromucous glands.

3.1.5.5
IgA Immunostaining in "Inflamed" (Surgically Obtained) Nasolacrimal Systems

Immunoreactive plasma cells were visible in the lamina propria of cases revealing moderate forms of chronic fibrosis. In the final states of dacryostenosis, plasma cells were only detected in front of or behind the stenotic area.

Conclusions (3.1.5). Organised lymphoid tissue is found with the cytomorphological and immunophenotypic features of MALT in nearly one third of the cases examined.

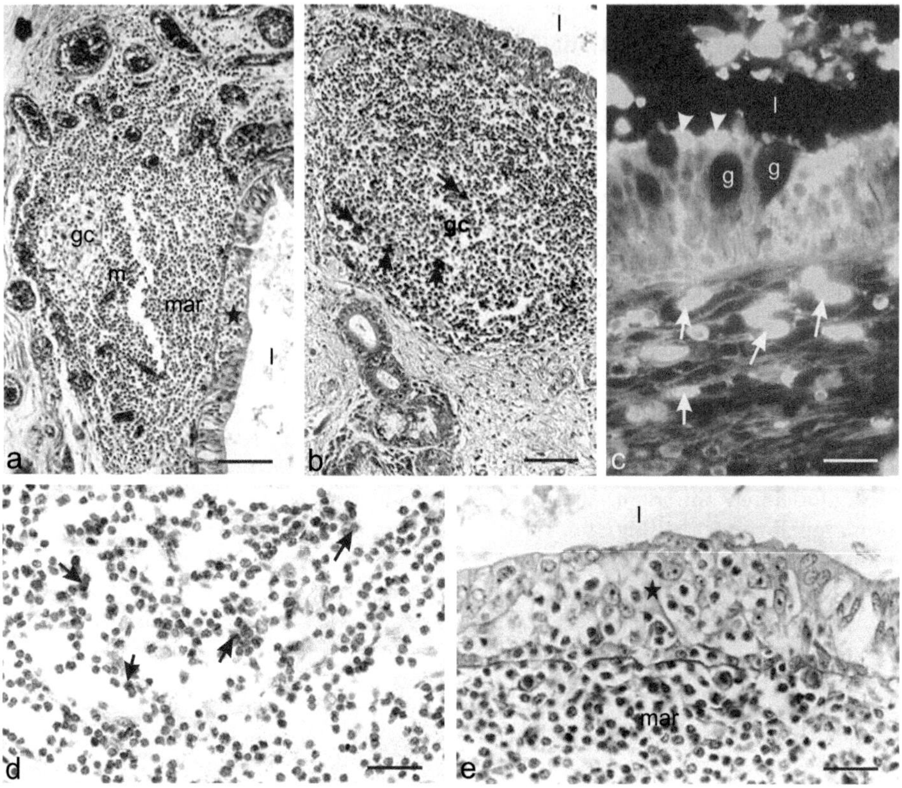

Fig. 12a–e. Immunohistochemistry of mucosa-associated lymphoid tissue. **a** Tear duct-associated lymphoid tissue (TALT) in a patient (female, 68 years) at the area of the prestenotic rim. TALT of the lacrimal sac with a well-developed germinal centre (*gc*), a mantle zone (*m*), a marginal zone (*mar*), and an overlying lymphoepithelium (*star*). *l*, Lacrimal passage; azan staining; *bar*=100 µm. **b** TALT in a cadaver (female, 67 years) with a well-developed germinal centre (*gc*), a surrounding mantle zone (*m*), a marginal zone (*mar*), and an overlying lymphoepithelium (*star*). Macrophages (*arrows*) are stained red by KiM1p. *l*, Lacrimal passage; immunostain – anti-CD68; *bar*=100 µm. **c** Immunohistochemical evidence of IgA in the lacrimal sac of a cadaver (female, 83 years). IgA-positive plasma cells (*arrows*) are located subepithelially. The epithelial cells show weak positivity for IgA, which is also contained in the mucous layer (*arrowheads*) overlying the epithelium. *g*, Goblet cell; *l*, lacrimal passage; immunostain – anti-IgA; *bar*=27.5 µm. **d** Anti-CD3 immunostaining shows red staining of T lymphocytes (*arrows*) within a lymphoid follicle of the lacrimal sac of a patient (female, 61 years). Immunostain – anti-CD3; *bar*=52 µm. **e** Frontal section through the lining epithelium of the nasolacrimal duct of a patient (female, 66 years) in the area of the poststenotic rim showing subepithelial parts of the marginal zone (*mar*) of a lymphoid follicle. The epithelium shows features of a lymphoepithelium (*star*) containing clusters of intraepithelial lymphocytes with irregular nuclei. *l*, Lacrimal passage; toluidine blue staining. *Bar*=46 µm

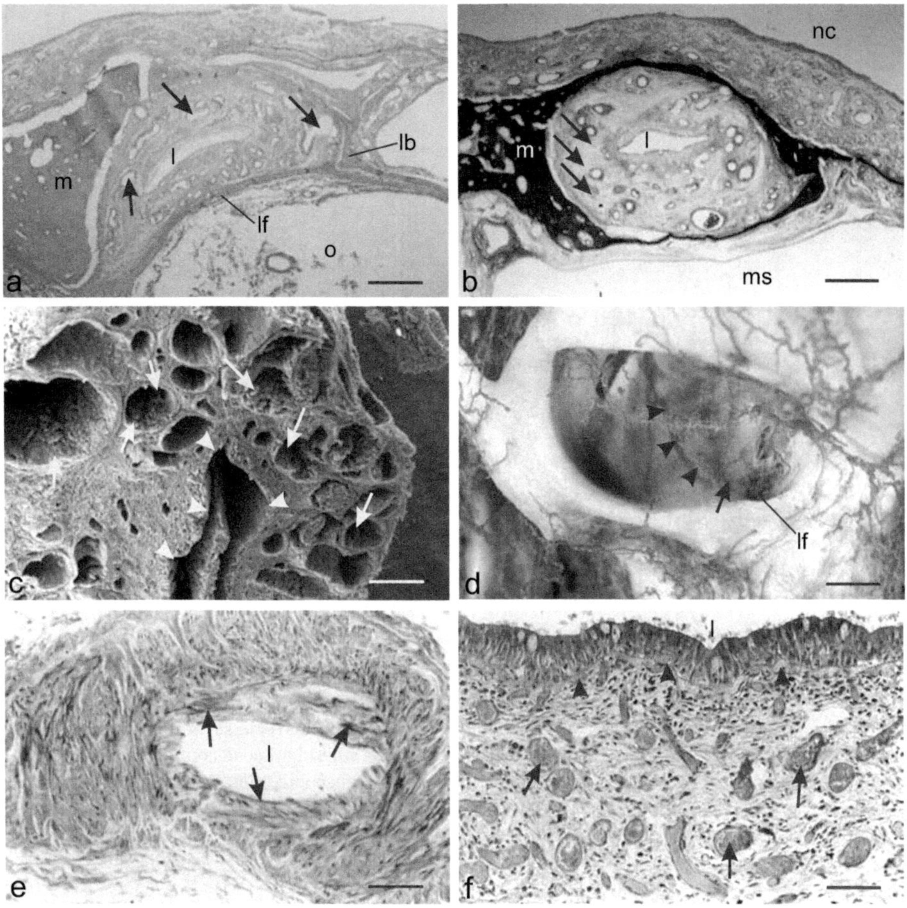

Fig. 13a–f. The vascular system of the nasolacrimal ducts. **a** Cross-section through the lower part of the lacrimal sac in a resorcin-fuchsine-thiazine picric-acid staining. More than two-thirds of the surrounding bony canal is filled by vascular plexus (*arrows*). *m*, Maxillary bone; *lb*, lacrimal bone; *l*, lumen of the lacrimal sac; *lf*, lacrimal fascia; *o*, orbit after enucleation; *bar*=2.6 mm. **b** Cross-section through the nasolacrimal duct with azan staining. More than two-thirds of the surrounding bony canal is filled by vascular plexus (*arrows*). *m*, Maxillary bone; *ol*, os lacrimale; *l*, lumen of the nasolacrimal duct; *nc*, nasal cavity; *ms*, maxillary sinus; *bar*=2.6 mm. **c** Scanning electron microscopic photograph of a horizontally sectioned lacrimal system. Wide-luminated blood vessels (*arrows*) surround the lumen of the lacrimal passage (*arrowheads*). *Bar*=1.3 mm. **d** Corrosion vascular cast of a right orbit. *Arrowheads* mark the ophthalmic artery with a branch (*arrow*) to the lacrimal fossa (*lf*). *Bar*=1 cm. **e** Transverse section of an artery with Goldner staining. Luminally, the arterial wall consists of an additional layer of longitudinally arranged smooth muscle cells (*arrows*). *l*, Lumen of the artery; *bar*=88 µm. **f** Cross-section through the subepithelial connective tissue of the lacrimal sac. Blood from a subepithelially located capillary network (*arrowheads*) is collected by post-capillary venules (*arrows*). *l*, Lumen of the lacrimal sac; toluidine blue staining; *bar*=88 µm

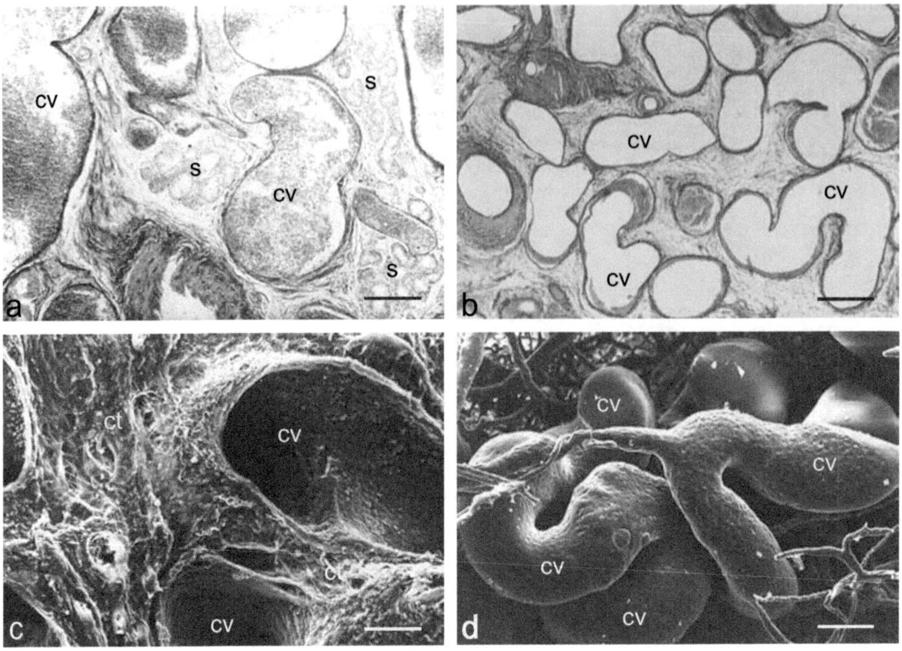

Fig. 14a–d. Capacitance veins. **a** Cross-section of some convoluted venous lacunae (*cv*) which are called capacitance veins. Erythrocytes are visible in the lumen of the veins. Some serous glands (*s*) are localised between the veins. Azan staining; *bar*=175 μm. **b** Transverse section of convoluted venous lacunae (*cv*) in a resorcin-fuchsine-thiazine picric-acid staining. Loose connective tissue is visible between the veins. *Bar*=345 μm. **c** Scanning electron microscopic photograph of convoluted venous lacunae (*cv*). Their lumens are opened by maceration with HClO. *ct*, Connective tissue between the blood vessels; *Bar*=147 μm. **d** Scanning electron microscopic photograph of a corrosion vascular cast showing convoluted venous lacunae (*cv*). *Bar*=222 μm

These findings suggest that tear duct-associated lymphoid tissue (TALT) is a feature that, although it need not be present in normal efferent tear ducts, is acquired during life in a proportion of apparently asymptomatic individuals. As TALT loss is associated with the scarring of symptomatic dacryostenosis, it is unlikely that the presence per se of TALT leads to scarring.

3.2 The Membranous Part of the Nasolacrimal Ducts

3.2.1 The Vascular System

The lacrimal sac and nasolacrimal duct were surrounded by a vascular plexus connected to the cavernous body of the inferior turbinate (Fig. 13a–c). The sac, flattened above but more rounded where it joins the duct, was enclosed in an osseofibrous cavern formed by the lacrimal fascia bridging the lacrimal fossa (Fig. 13a). The duct

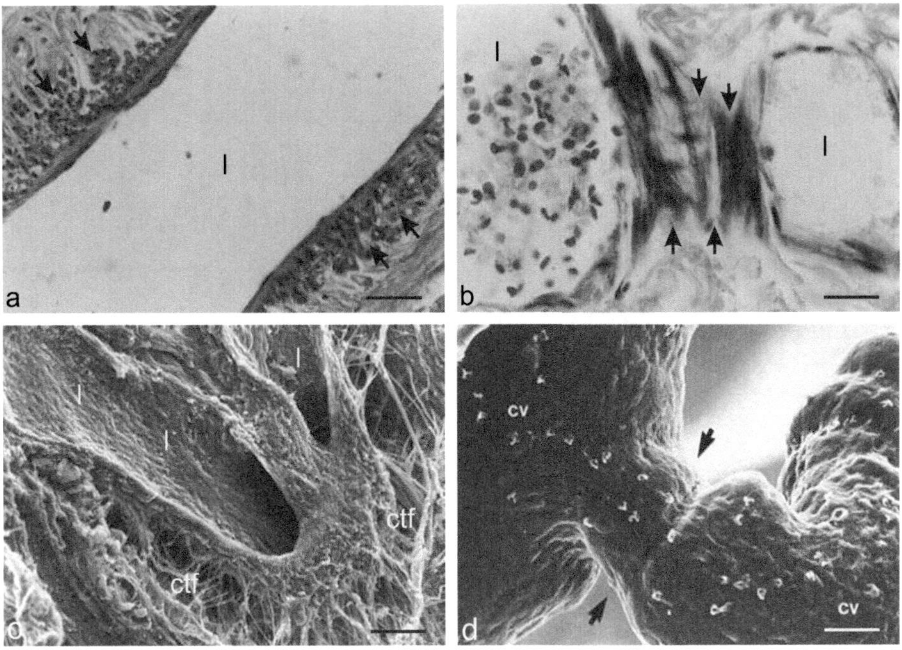

Fig. 15a–d. Throttle veins. **a** Cross-section of a venous lacuna. *Arrows* mark muscle fibres of a markedly developed tunica media. *l*, Lumen of the blood vessel. Goldner staining; *bar*=44 µm. **b** Transverse section of a venous lacuna. *Arrows* mark muscle fibres which are circularly arranged around the lumen of this so-called throttle vein. *l*, Lumen of venous lacuna; Goldner staining; *bar*=28 µm. **c** Scanning electron microscopic photograph of a specialised (throttle) vein. In the wall of the vein numerous recesses are localised between a network of connective tissue fibres (*ctf*) in which normally smooth muscle cells are embedded. The muscle cells have been removed by a maceration process. *l*, Lumen of the throttle vein; *bar*=105 µm. **d** Scanning electron microscopic photograph of a corrosion vascular cast of a venous lacuna (*cv*) or so-called capacitance vein. The lumen of the vein is narrowed in its middle segment (*arrows*). Such a segment is termed a throttle vein. *Bar*=32 µm

was embedded in a bony canal formed by the maxilla and the lacrimal bone (Fig. 13b). More than two-thirds of the bony canal between orbit and inferior turbinate were occupied by the vascular plexus (Fig. 13b). The main blood supply of the efferent tear ducts derived from the ophthalmic arteries (Fig. 13d) and to a lesser extent from the infraorbital and sphenopalatine arteries. Their branches ran in a craniocaudal-to-caudocranial direction through the lacrimal passage.

Subepithelially, the lamina propria consisted of collagen bundles as well as elastic and reticular fibres arranged in a helical pattern and encloses some mixed glands (Fig. 2h) with excretory canals that opened at the surface of the epithelium.

Comparison between the findings made by light microscopy, corrosion vascular casts as well as scanning electron microscopy revealed different blood vessel types inside the vascular system of the nasolacrimal ducts.

Thick-walled, muscular arteries located near the periosteum, or in contact with it, gave off branches that ran vertically through the lamina propria. Segments of these

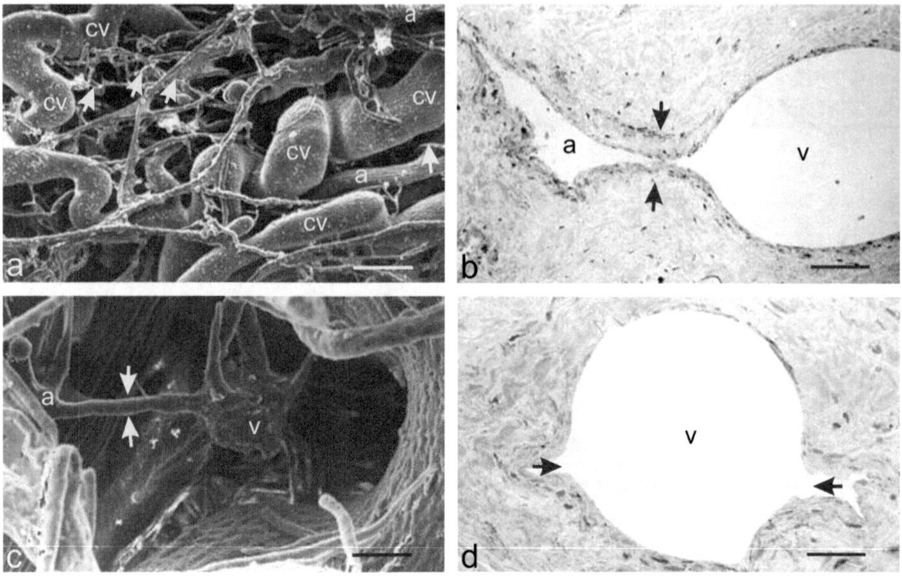

Fig. 16a–d. Arteriovenous anastomoses. **a** Scanning electron microscopic photograph of a corrosion vascular cast. Numerous arteriovenous anastomoses (*arrows*) showing the form of short bridges are localised between branches of arteries (*a*) and convoluted venous lacunae (*cv*). *Bar*=270 µm. **b** Transverse section of an arterio (*a*) – venous (*v*) anastomosis (*arrows*). Toluidine blue staining; *bar*=88 µm. **c** Arterio (*a*) – venous (*v*) anastomosis (*arrows*) demonstrating the form of a short bridge. *Bar*=102 µm. **d** Cross-section through a vein with two orifices of arteriovenous anastomoses (*arrows*). Toluidine blue staining; *bar*=88 µm

arteries consisted of an additional layer of longitudinally arranged smooth muscle cells located luminally to the usual layer of helically arranged smooth muscle cells (Fig. 13e). The branches divided just beneath the epithelium into superficial arcading branches (Fig. 13f). The blood from the capillary network was collected by short postcapillary venules that drained into widely convoluted venous lacunae (Fig. 14a–d). The diameter between the lacunae varied between 0.2 and 0.6 mm. In most cases, the tunica intima contained a thin subendothelial layer, whereas the tunica media was sparingly developed, and the adventitia was clearly visible as a broad band of connective tissue. Valves were not seen inside the lacunae.

At numerous places, the lacunae consisted of a markedly developed musculature (Fig. 15a–d). Venous lacunae were connected to veins situated near bone. Some veins of the vascular plexus were situated close to the wall of the bony canal and were connected to intraosseous veins of the maxilla or the lacrimal bone.

Between the arteries and veins were numerous arteriovenous anastomoses (Fig. 16a–d) characterised by the presence of epithelioid cells. The arteriovenous anastomoses had the form of short bridges, and no tortuous or glomerular anastomoses were found in the efferent tear ducts (Fig. 16a–d).

Conclusions (3.2.1). The surrounding vascular plexus of lacrimal sac and nasolacrimal duct is comparable to a cavernous body. While, regulating the blood flow, the specialised blood vessels permit opening and closing of the lumen of the lacrimal passage effected by the bulging and subsiding of the cavernous body, at the same time regulating tear outflow. Other functions such as drainage of absorbed tear fluid components, temperature exchange and a role in immunologic response are under discussion as well. Malfunctions in the cavernous body may lead to disturbances in the tear outflow cycle, ocular congestion or total occlusion of the lacrimal passages. Variations in the conditions for swelling of the cavernous tissue may have led to the (mistaken) description of valves in the lacrimal passage.

3.2.2 Innervation of the Vascular System and the Seromucous Glands of the Nasolacrimal Ducts

To evaluate the innervation of the arteries and veins in the cavernous body of the human efferent tear ducts, along with its surrounding structures including small seromucous glands (Fig. 2h) the ultrastructure and distribution of nerve fibres immunoreactive with antisera against neurofilaments (NSE, S-100, RT-97), neuropeptides (SP, NPY, CGRP, VIP) and neuronal enzymes (TH) were examined by immunohistochemistry and TEM.

The blood vessels of the cavernous body of the efferent tear ducts were supplied by a dense innervation. Subendothelially, myelinated as well as non-myelinated nerve fibres were found along arteries and veins that had a pronounced tunica media. Profiles of nervous tissue were also numerous in seromucous glands.

3.2.2.1 Distribution of Neuronal Markers

200-kDa Neurofilament. RT97-immunoreactive myelinated thicker nerve fibre bundles were found loosely distributed throughout the bulk of connective tissue of the subepithelial layer and between the blood vessels of the cavernous body (Fig. 17a).

S-100 Protein (The Antiserum Reacting with Both S-100α and S-100β Subunits). S-100 immunoreactivity occurred in about the same density as the RT97-immunoreactive nerve fibres (Fig. 17c). Apart from these coarse nerve fibre bundles, immunofluorescence of S-100 was visible in the area of the mixed glands, the subepithelial layer and in the mucosa. S-100 immunoreactive nerve fibres were also seen to enter the walls of interlobular blood vessels. In addition, within the parenchyma of mixed glands, scattered groups of epithelial cells were S-100 immunoreactive, while the majority of the cells remained unstained.

Neuron-Specific Enolase. NSE-immunoreactive structures were identified in the subepithelial region, around the blood vessels of the cavernous body and also abundantly in the area of the seromucous glands. Moreover, some fine nerve fibres formed a network around secretory cells of the seromucous glands.

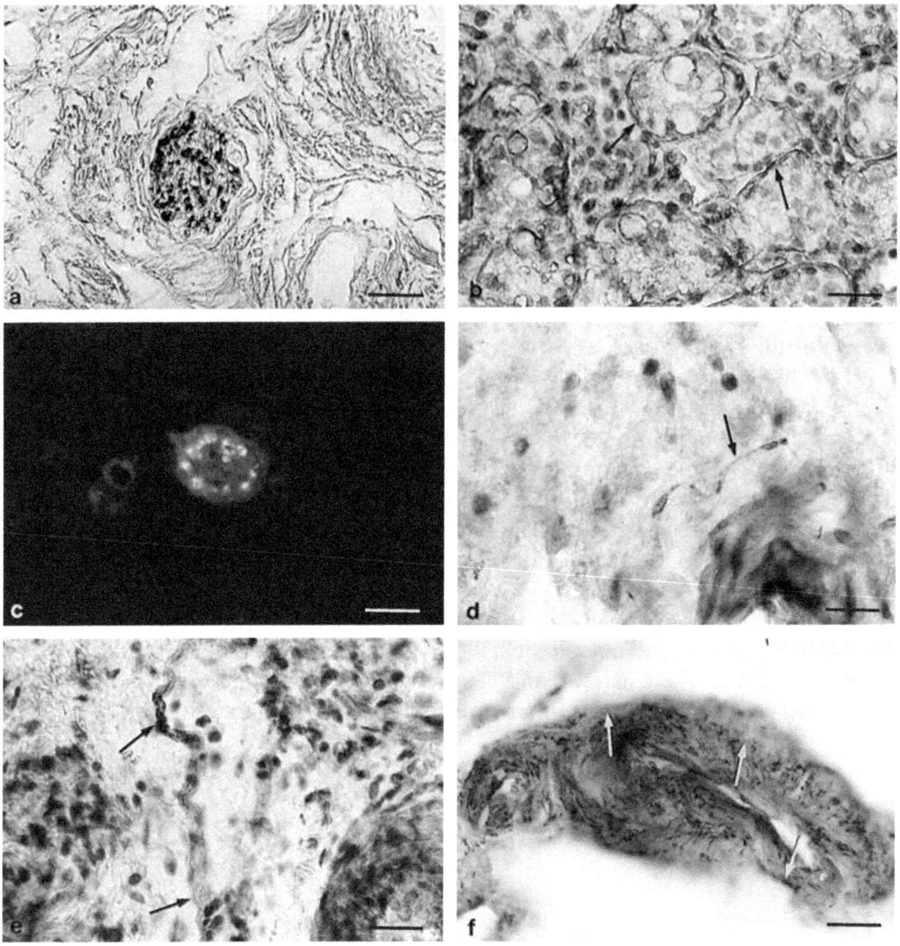

Fig. 17a–f. Distribution of nervous tissue and neuropeptides. **a** Visualisation of a myelinated nerve by immunoreaction with RT-97 in the connective tissue of the cavernous body. *Bar*=41 μm. **b** VIP-immunoreactive nerve fibres (*arrows*) in association with secretory tubules of serous glands. *Bar*=26 μm. **c** Distribution of S-100 protein in a myelinated nerve of the cavernous body. *Bar*=26 μm. **d** SP-immunoreactive nerve fibres (*arrows*) in association with a blood vessel of the cavernous body. *Bar*=26 μm. **e** Visualisation of a TH-immunoreactive nerve fibre (*arrows*) in the connective tissue of the cavernous body. *Bar*=27 μm. **f** Distribution of TH-immunoreactive nerve fibres (*arrows*) in the wall of a barrier artery. *Bar*=83 μm

Vasoactive Intestinal Polypeptide. VIP-immunoreactive structures were only stained within the body of seromucous glands of the efferent lacrimal tear ducts. Between the secretory tubules, scattered portions of fine and also of thicker nerve fibres were stained (Fig. 17b). Fibres around blood vessels in the intralobular and interlobular connective tissue septa also appeared to show a positive reaction to the antibody.

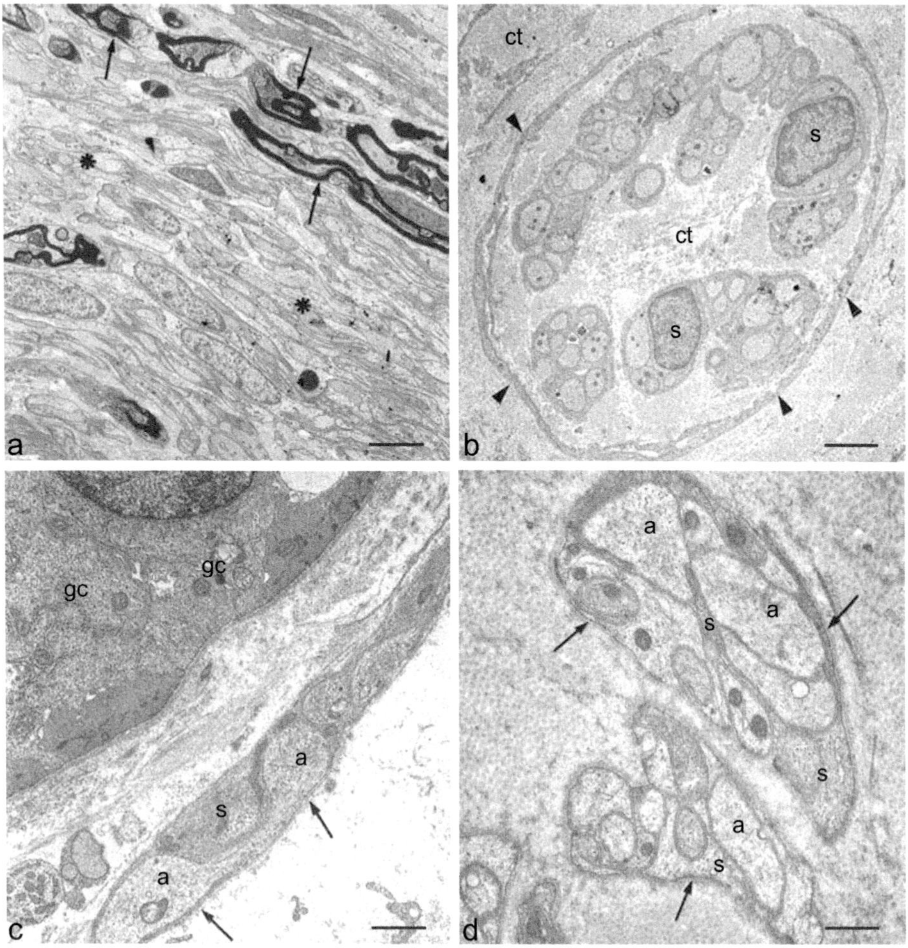

Fig. 18a–d. TEM of nervous tissue. **a** Longitudinal section through a thick nerve bundle inside the connective tissue of the cavernous body. In addition to myelinated nerve fibres (*arrows*), many unmyelinated nerve fibres (*stars*) are visible. *Bar*=9 μm. **b** Cross-section of a small unmyelinated nerve surrounded by perineurium in the connective tissue of the cavernous body. The perineurium (*arrowheads*) completely surrounds the endoneurium, which contains numerous unmyelinated axons and associated Schwann cells (*s*). The latter are separated from the endoneurial connective tissue (*ct*) by basement membranes. *Bar*=2.5 μm. **c** Part of an unmyelinated nerve fibre (*arrows*) in close association with serous acinar cells (*gc*). *a*, Axon; *s*, cytoplasm of Schwann cell; *bar*=1.6 μm. **d** Unmyelinated nerve fibres (*arrows*) in the connective tissue (*ct*) of a capacitance vein. *a*, Axon; *s*, Schwann cell cytoplasm; *bar*=7.5 μm

Calcitonin Gene-Related Peptide. Few calcitonin gene-related peptide (CGRP)-immunoreactive nerve fibres were identified in the intertubular stroma. The stained fibres were associated with secretory tubules, intralobular ducts and blood vessels. Around the acinar cells of the glands thicker nerve fibre bundles were stained.

Neuropeptide Y. NPY-immunoreactivity was associated with blood vessels of the cavernous body, the subepithelial capillary bed and the blood vessels of the seromucous glands. The main localisation of NPY-immunoreactive nerve fibres was around and in the wall of arterioles, where fine positive fibres were identified intermingled with smooth muscle cells of the tunica media.

Tyrosin Hydroxylase. TH-immunoreactive nerve fibres were identified loosely scattered throughout the connective tissue of the efferent lacrimal tear ducts (Fig. 17e). They were also detected in abundance between the lobules of seromucous glands and in association with blood vessels of the cavernous body (Fig. 17f).

Substance P. SP-immunoreactive nerve fibres were seen in a distinct association with glandular epithelium of seromucous glands as well as with blood vessels inside the lobules of the seromucous glands, and the vascular system of the cavernous body (Fig. 17d). Thicker nerve fibre bundles were stained around the seromucous glands and blood vessels.

3.2.2.2 Ultrastructure of the Nervous Tissue

Nerve bundles composed of myelinated and unmyelinated fibres (Fig. 18a) were visible in the connective tissue of the subepithelial layer and the cavernous body.

Profiles of nervous tissue were also numerous in seromucous glands. The nerve fibres were surrounded by a perineurial sheath (Fig. 18b). This enclosed either single Remak fibres with multiple axons or a combination of myelinated and unmyelinated fibres. The endoneurial spaces contained collagen fibrils (Fig. 18b). Numerous profiles of nerve fibres showed variable numbers of axons free of the surrounding Schwann cell cytoplasm within a defined area. Such nerve fibres were located immediately adjacent to the basal cell layers of the alveoli in narrow connective tissue septa between alveolar buds (Fig. 18c) and in the neighbourhood of capillaries. The axons form so-called "synapses á distance" with basally located glandular cells and with blood vessels. The axons contained small, clear vesicles and larger, dense core vesicles.

The blood vessels of the cavernous body of the efferent lacrimal tear ducts were also supplied by a dense innervation. Unmyelinated, longitudinally cut nerve bundles and smaller axon conglomerates were detected in the arterial adventitia (Fig. 18d). Generally, in contrast to arteries, veins showed few nerve structures which, again in contrast to the arteries, were also located between the smooth muscle cells. However, some of the veins displayed a more pronounced innervation pattern, particularly in their muscular component. In contrast to smaller arterioles, no axons were demonstrated adjacent to capillary walls.

Conclusions (3.2.2). Both the density of nerve fibres as well as the presence of various neuropeptides emphasises the neural control of the cavernous body of the human efferent tear ducts. By means of this innervation, the specialised blood vessels permit regulation of blood flow by opening and closing of the lumen of the lacrimal passage as effected by the engorgement and subsidence of the cavernous body, at the

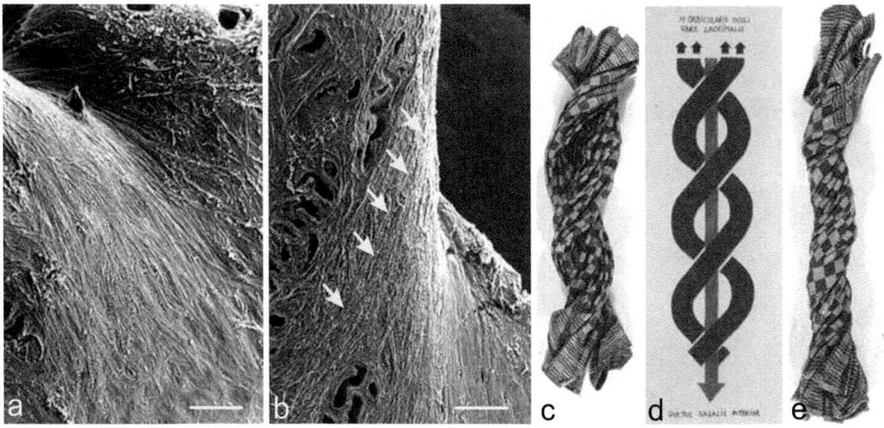

Fig. 19a–e. Fibrillar structures of the nasolacrimal ducts. **a** SEM photograph of the outer surface of a nasolacrimal duct revealing the helical arrangement of collagen fibrils. Note the "screw-shaped" arrangement of the collagen fibrils. *Bar*=550 µm. **b** SEM photograph of the outer surface of a nasolacrimal duct revealing the helical arrangement of collagen fibrils (*arrows*). *Bar*=550 µm. **c–e** The arrangement of collagen and elastic fibres in a schematic view. If the nasolacrimal duct distends while blinking it will be "wrung out" due to the screw-shaped arrangement of the collagen bundles

same time regulating tear outflow. Related functions such as a role in the occurrence of epiphora related to emotional responses are relevant. Moreover, malfunctions in the innervation of the cavernous body may lead to disturbances in the tear outflow cycle, ocular congestion, or total occlusion of the lacrimal passages.

3.2.3 Structural Analysis of the Lamina Propria in the Nasolacrimal Ducts

The lacrimal sac was attached medially in the lacrimal fossa. Fibroelastic and muscular elements of attachment were situated rostral to the fornix. In this region the lacrimal sac was attached to the lacrimal part of the orbicularis oculi muscle by fibrillar structures. Caudally, at the transition to the inferior meatus of the nose, the nasolacrimal duct was fixed by collagen fibrils at the inferior turbinate.

Between the epithelial lining of the lacrimal sac and the nasolacrimal duct, the above-described cavernous body was situated. Caudally it was connected to the "cavernous body" of the inferior turbinate. More than two-thirds of the bony canal between orbit and inferior turbinate was filled by the cavernous body (Fig. 13a–c).

The connective tissue between the blood vessels of the cavernous body was made up of collagen bundles, elastic and reticular fibres arranged in a helical pattern. In this helical system, fibres ran screw-shaped from the fornix to the outlet of the nasolacrimal duct (Fig. 19a, b). Immunohistochemical analysis confirmed the presence of collagen type I and type III as well as chondroitin-4-sulphate, chondroitin-6-sulphate and keratan sulphate as main components of the extracellular matrix in the wall of the efferent tear duct system.

Conclusions (3.2.3). With blinking, the lacrimal part of the orbicularis muscle contracts. The fornix of the sac moves in a cranial-lateral direction. Thus, the lacrimal sac distends and may be "wrung out" due to its medial attachment and helically arranged fibrillar structures. At the same time, lacrimal fluid inside the lumen of the lacrimal passage is drained distally (Fig. 19c–e).

3.3 Absorption Experiments in the Nasolacrimal Ducts

3.3.1 Absorption Experiments in Rats

3.3.1.1 Kinetics of Uptake of Radioactivity into the Serum

Radioactivity detected in the serum was low, but increased with time. After 10 min 27 Bq/ml, after 20 min 36 Bq/ml and after 60 min 247 Bq/ml were measured. This corresponded to an incorporation of maximal 0.13% of the applied radioactivity into the blood assuming a total blood volume of 10 ml.

3.3.1.2 Analysis of the Incorporated Radioactivity in the Serum

Rough analysis by TCA precipitation showed that after all times most of the radioactivity (70%–80%) was not protein-bound. This was further analysed and could be sustained by gel chromatographic separation of the serum according to molecular mass. Three radioactivity peaks were fractionated: one corresponding to BSA with maximal 4.8 Bq/ml (after 60 min), one corresponding to iodinated tyrosine (5.5 Bq/ml after 60 min), and one corresponding to free iodine (237 Bq/ml after 60 min). This showed that only traces of the iodinated BSA applied were incorporated into the circulation. An even higher part of the radioactivity was taken up as the proteolytic degradation product of BSA to free amino acids, and 96% of the radioactivity incorporated was free iodine.

3.3.1.3 Light Microscopy and Autoradiography

The efferent tear ducts of the rat consisted of a tube starting with an aperture at the ventral-caudal part of the upper palpebral conjunctiva (Fig. 20a, b). From here it ran in an anterio-medio-caudal direction and opened below the posterior part of the inferior turbinate into the nose. The tube was lined by a multilayered epithelium (Fig. 20c). The upper cell layers consisted of larger squamous elements over several layers of more or less cuboidal cells. In many places, compact intraepithelial nests of clear cells were integrated. They formed characteristic intraepithelial mucous glands

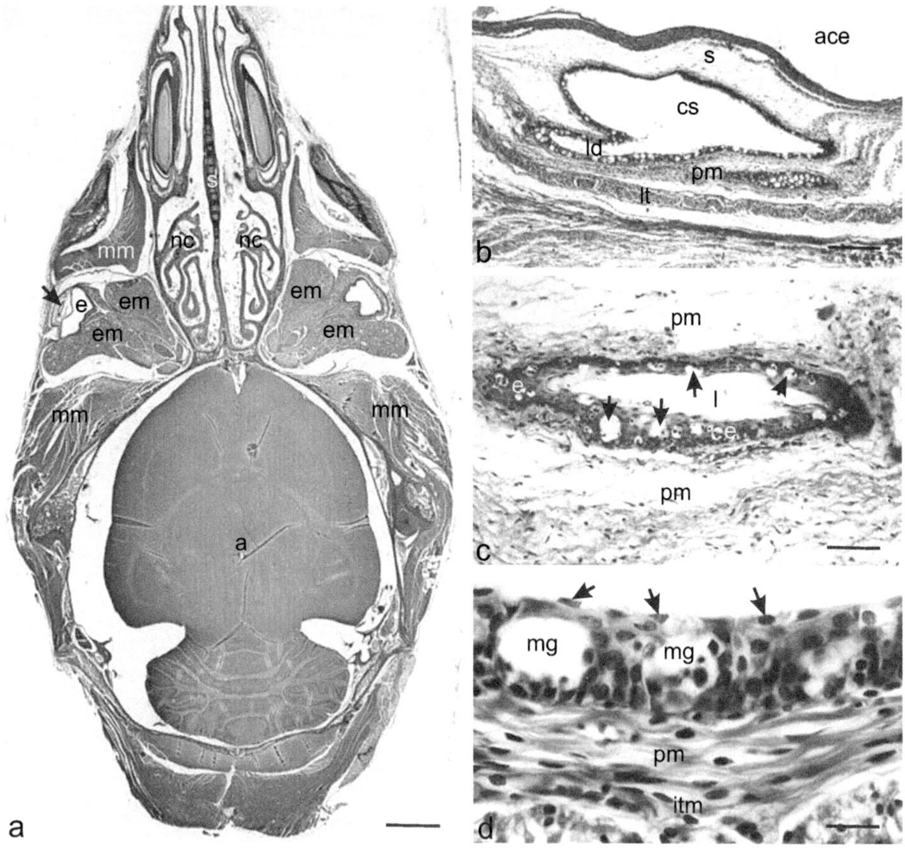

Fig. 20a–d. Horizontal section through a rat head. *Arrow* marks efferent tear duct near the eye (*e*). *em*, Eye musculature; *mm*, masticatory musculature; *nc*, nasal cavity, *s*, nasal septum; *c*, cerebrum; *bar*=2.2 mm. **b** *Arrow* marks beginning of the nasolacrimal duct (*ld*) at the conjunctival sac. *s*, Sclera; *ace*, aqueous chamber of the eye; *pm*, propria mucosae; *it*, inferior tarsal muscle; *bar*=351 μm. **c** Horizontal section through the nasolacrimal duct in higher magnification. *l*, Lumen of the duct; *e*, epithelium with integrated mucosal glands (*arrows*); *pm*, propria mucosae; *bar*=176 μm. **d** Higher magnification of the lining epithelium of the nasolacrimal duct with large squamous cells at the epithelial surface (*arrows*) and cuboid cells in deeper epithelial layers. *mg*, Intraepithelial mucosal glands; *pm*, propria mucosae; *itm*, inferior tarsal muscle; *bar*=28 μm

(Fig. 20d). The secretory product of these cells reacted strongly positive with alcian blue (pH 1) and with mPAS. Underneath the epithelium, the lamina propria consisted of loose connective tissue containing a thin layer of elastic fibres and sometimes lymphatic cells (Fig. 20c). Additionally, the lamina propria contained some large veins near the opening of the nasolacrimal duct into the nose. After autoradiography, none of these sections in different horizontal planes showed any blackening of the X-ray film, independent of exposure time.

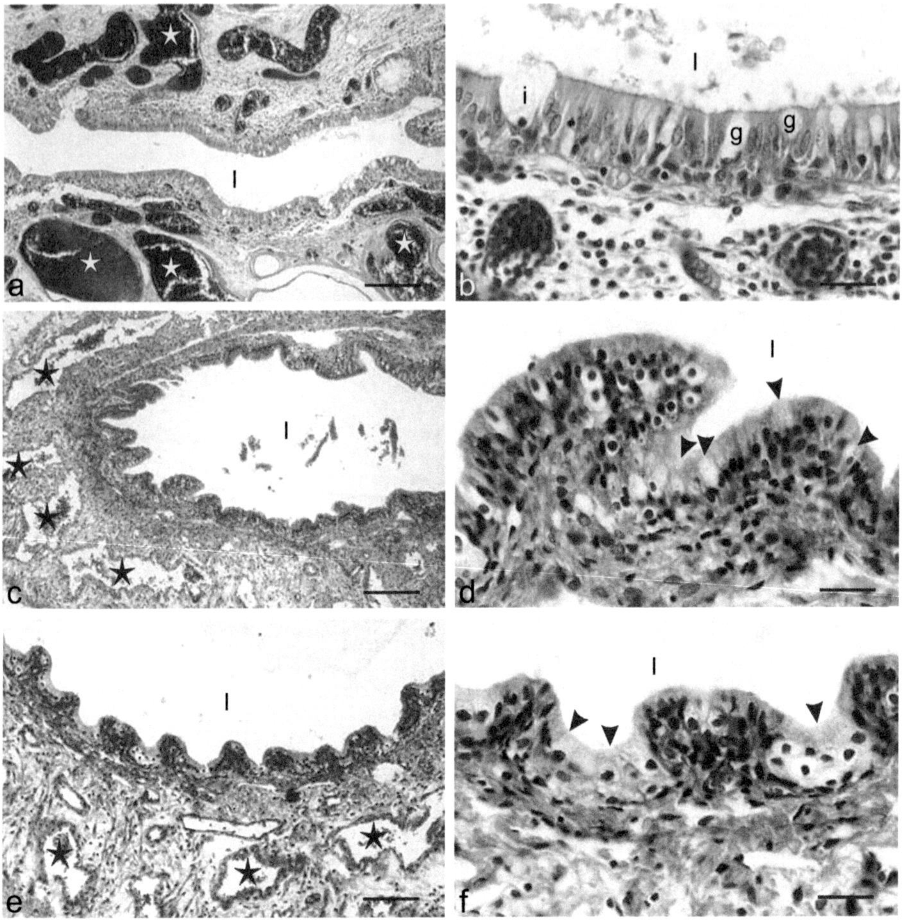

Fig. 21a–f. Comparative anatomy of the nasolacrimal ducts. **a** Horizontal section of a human lacrimal sac. The lumen of the lacrimal passage is surrounded by a cavernous body rich in blood vessels with wide lumina (*stars*). *Bar*=170 µm. **b** Higher magnification of **a**. The lacrimal sac contains a double-layered epithelium. Goblet cells are integrated as solitary cells (*g*) or show a characteristic arrangement of several cell groups forming mucous glands (*i*). *Bar*=27 µm. **c** Horizontal section through the nasolacrimal duct of an ape. The lumen of the lacrimal passage is surrounded by a cavernous body rich in blood vessels with wide lumina (*stars*). *Bar*=170 µm. **d** Higher magnification of **c** revealing the epithelium. Some of the epithelial cells reveal a goblet cell-like appearance with pale supranuclear cytoplasm (*arrowheads*). *Bar*=27 µm. **e** Horizontal section through the nasolacrimal duct of a rabbit. The lumen of the lacrimal passage is surrounded by a cavernous body rich in blood vessels with wide lumina (*stars*). *Bar*=85 µm. **f** Higher magnification of **e**, revealing the epithelium. Some of the epithelial cells reveal a goblet cell-like appearance with pale supranuclear cytoplasm (*arrowheads*). These cells are mostly arranged in cell groups. *Bar*=27 µm. *a–f*, Goldner staining; *l*, lumen of the lacrimal passage

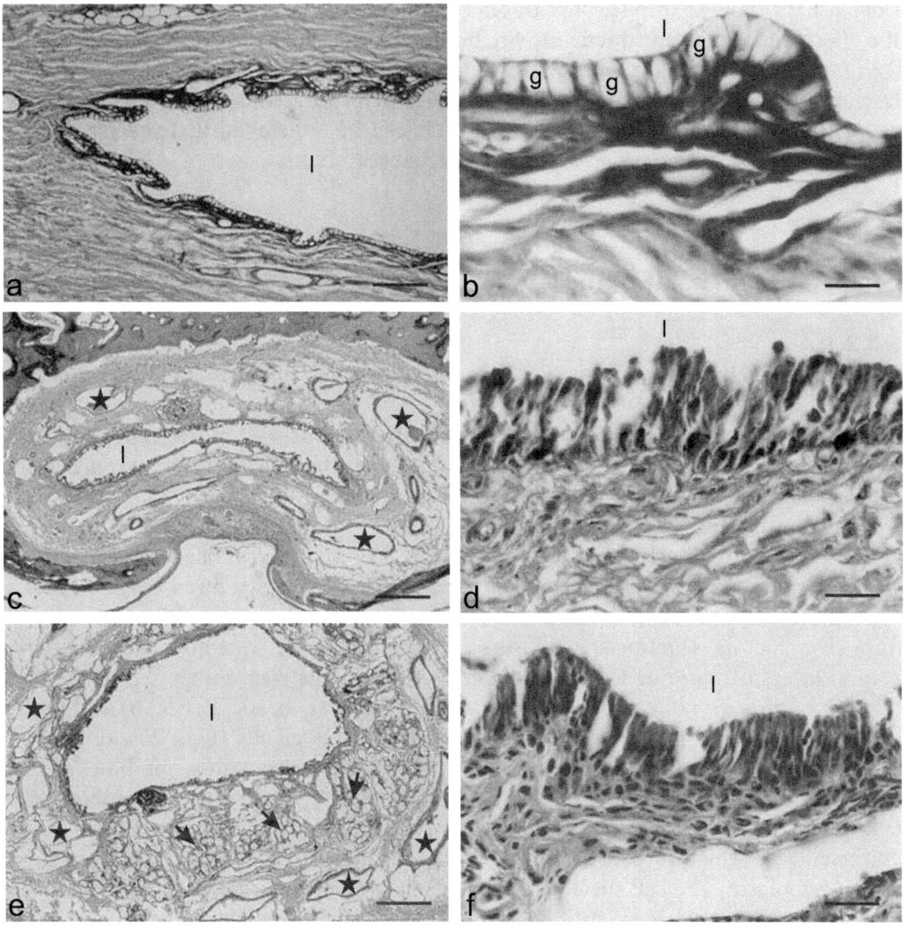

Fig. 22a–f. Comparative anatomy of the nasolacrimal ducts. **a** Horizontal section of a nasolacrimal duct from a cat. The lumen is surrounded by strong connective tissue in which some blood vessels are visible. However, a surrounding cavernous body is absent. *Bar*=170 µm. **b** Higher magnification of **a** revealing the lining epithelium of the nasolacrimal duct. The pseudo-stratified columnar epithelium contains many goblet cells (*g*). *Bar*=27 µm. **c** Horizontal section through the nasolacrimal duct of a deer. The lumen of the lacrimal passage is surrounded by a cavernous body rich in blood vessels with wide lumina (*stars*). *Bar*=333 µm. **d** Higher magnification of **c**, revealing a pseudostratified columnar epithelium without goblet cells. *Bar*=27 µm. **e** Horizontal section through the nasolacrimal duct of a pig. The lumen of the lacrimal passage is surrounded by a cavernous body rich in blood vessels with wide lumina (*stars*). The lamina propria contains many seromucous glands (*arrows*). *Bar*=333 µm. **f** Higher magnification of **e**. The duct contains a pseudostratified columnar epithelium without goblet cells. *Bar*=27 µm. **a–f**, Goldner staining; *l*, lumen of the lacrimal passage

Conclusions (3.3.1). In rats, only traces of iodinated albumin are incorporated from the efferent lacrimal tear ducts into the blood. An even higher part of the radioactivity is taken up as the proteolytic degradation product of bovine serum albumin to free amino acids, and 96% of the radioactivity incorporated is free iodine, probably as a contaminant of iodinated preparation. It can be concluded that the rat is unfit for studying absorptive processes in the efferent tear ducts.

3.3.2 Comparative Anatomy of the Nasolacrimal Ducts

3.3.2.1 Comparative Histology

Except for the human and ape nasolacrimal systems composed of the upper and lower canaliculi, the lacrimal sac, and the nasolacrimal duct, all lacrimal systems from the vertebrates investigated consisted solely of the upper and lower canaliculus leading directly into the nasolacrimal duct.

In light microscopy, human, ape, rabbit, cat, deer and pig revealed a pseudostratified, columnar epithelium with double layering in most areas, a basal cell layer and a superficial columnar layer (Figs. 21a–f, 22a–f). The rat showed a multilayered epithelium (Fig. 20c, d). The upper cell layers consisted of larger squamous elements over several layers of more or less cuboidal cells. Goblet cells were integrated in the epithelium of human, rat, and cat as solitary cells (Figs. 21a, b, 20c, d, 22a, b) and in the epithelium of human, and rat as intraepithelial mucous glands (Figs. 20c, d, 21a, b). The secretory product of these cells reacted strongly positive with alcian blue (pH 1). By contrast, the epithelium of ape, rabbit, deer and pig contained no goblet cells (Fig. 21c–f, 22c–f). However, the epithelium of ape, rabbit, and pig contained many epithelial cells which revealed a soft positive staining with alcian blue (pH 1) in their upper cytoplasm. The cell surface reacted strongly positive with alcian blue. The cells with the soft staining were mostly arranged in cell groups (Fig. 21d,f). There were also epithelial areas without such cells or cell groups. Subepithelially, the lamina propria of the human lacrimal passage was composed of two strata: (1) loose connective tissue containing elastic fibres and lymphatic cells as well as (2) a rich venous plexus comparable to a cavernous body (Fig. 21a, b). A surrounding cavernous system of blood vessels was also found in ape, rabbit, deer and pig (Figs. 21c, e, 22c, e), but was absent in rat and cat (Figs. 20c, 22a). Small seromucous glands opening their excretory ducts into the lacrimal passage were integrated in the lamina propria of human and pig. In the pigs, seromucous glands were distributed along the whole nasolacrimal duct (Fig. 22e), whereas in the human specimens the glands were present only in the lacrimal sac in small numbers. None of the other animals possessed seromucous glands. The results are summarised in Table 5.

Table 5. Comparison of nasolacrimal ducts of humans with those of various other vertebrates

Specimen	Epithelium	GC	IMG	SCB	SSMG
Human	Double-layered	Yes	Yes	Yes	Yes, but only small numbers in lacrimal sac
Ape	Double-layered	No, but mucin secreting epithelial cells	No, but mucin secreting epithelial cells organised in cell groups	Yes	No
Rabbit	Double-layered	No, but mucin secreting epithelial cells	No, but mucin secreting epithelial cells organised in cell groups	Yes	No
Rat	Multilayered	Yes	Yes	No	No
Cat	Double-layered	Yes, many	No	No	No
Deer	Double-layered	No	No	Yes	No
Pig	Double-layered	No, but mucin secreting epithelial cells	No	Yes	Yes, throughout entire nasolacrimal duct

GC, goblet cells; IMG, intraepithelial mucous glands; SCB, surrounding cavernous body; SSMG, subepithelial seromucous glands.

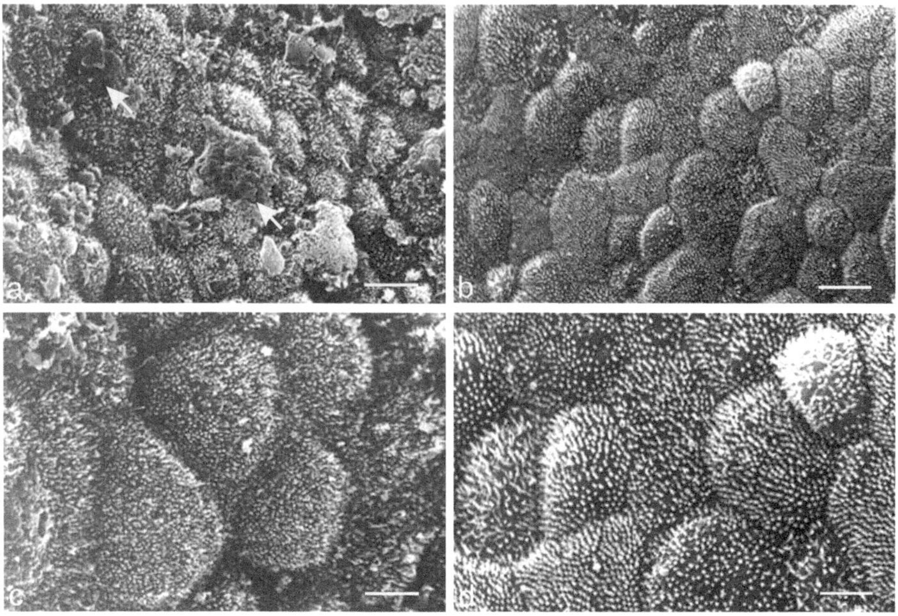

Fig. 23a–d. SEM micrographs of the surface of epithelial cells in the nasolacrimal ducts of human (**a**, **c**) and rabbit (**b**, **d**). Cell borders are clearly distinguishable in all figures. SEM reveals surface covering of epithelial cells consisting of a trimming with microvilli in both human and rabbit. *Arrows* mark secretory products of goblet cells (*a*). **c**, **d** Epithelial cells in higher magnification. Magnification: **a** and **b**, *bar*=10 μm; **c** and **d**, *bar*=5 μm

Fig. 24. Kinetics of uptake of radioactive ^{3}H-cortisol into blood/serum of rabbits. After 21 min, 41,470 disintegrations per minute (dpm)/ml, after 43 min 72,320 dpm/ml, and after 146 min 91,020 dpm/ml were measured

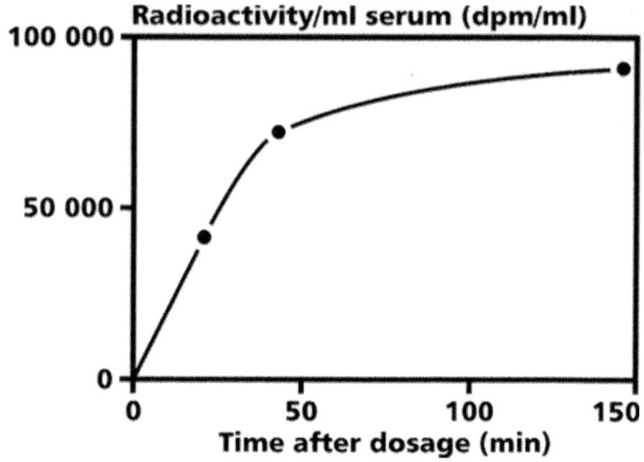

3.3.2.2
Comparative Scanning Electron Microscopy (Human and Rabbit)

Scanning electron microscopy (SEM) of human as well as rabbit nasolacrimal systems revealed the outer surface of the epithelial cells to be covered with microvilli (Fig. 23a–d). The length of the microvilli was between 400 and 600 nm (human) and 200–400 nm (rabbit), the diameter between 60 and 70 nm in both humans and rabbits. Goblet cells were clearly distinguishable from epithelial cells on account of their differing cell surface or secretions (Fig. 23a).

Conclusions (3.3.2). Besides primates, histology reveals the nasolacrimal systems of rabbits to be closely comparable to that of humans. It reveals a stratified epithelium consisting of two layers. Subepithelially, the lamina propria is composed of two strata: (1) loose connective tissue containing elastic fibres and lymphatic cells as well as (2) a rich venous plexus comparable to a cavernous body. Thus, rabbits are most suitable for studies regarding absorption inside the nasolacrimal ducts.

3.3.3
Absorption Experiments in Rabbits

3.3.3.1
Kinetics of Uptake of Radioactive Cortisol into Blood/Serum

Radioactivity detected in the serum increased in a time-dependent manner (Fig. 24). After only a short time, a readily measurable blood level was reached. After 21 min 7.1%, after 43 min 12.4%, and after 146 min/ml, 15.5% of the applied radioactivity was incorporated into the blood, assuming a total blood volume of 180 ml.

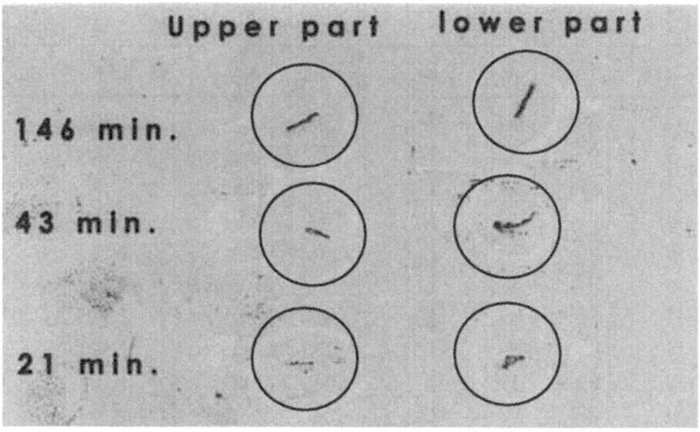

Fig. 25. Autoradiography of left rabbit nasolacrimal ducts after uptake of radioactive ^{3}H-cortisol. Nasolacrimal systems reveal a time-dependent blackening of the X-ray film

3.3.3.2
Light Microscopy and Autoradiography (Rabbit)

Light microscopy of the nasolacrimal ducts of laboratory rabbits (data not shown) used in the absorption experiments was comparable to that observed and already described above for wild rabbits in the comparative study. After autoradiography for 23 days, all six samples showed blackening of the X-ray film in a time depending manner. Tear duct samples of the 146-min rabbit revealed the clearest marks, followed by the 43- and then 21-min samples (Fig. 25). No differences were observed between the upper and lower segments of the nasolacrimal ducts except in the nasolacrimal duct of the 21-min rabbit. Here, the lower segment showed the least blackening (Fig. 25).

Conclusions (3.3.3). In rabbits, ^{3}H-cortisol is incorporated from the nasolacrimal ducts into the blood. Based on the comparable morphology of rabbits and humans, it is suggested that absorption of cortisol also takes place in humans. Future investigations in the nasolacrimal passage are needed in order to understand whether absorption of normal tear fluid components in the nasolacrimal ducts is a physiological function and also plays a role in pathological conditions, i.e. in dry eye. The similarities between rabbit and human nasolacrimal ducts support the use of the rabbit for such studies.

3.4
Pathologic Findings in Dacryostenosis

Thirty-six cases of clinically defined primary acquired dacryostenosis were selected (Table 6). Among these, 19 cases of stenosis were located in the nasolacrimal duct, 6 cases in the lower part of the lacrimal sac, and in 11 cases at the junction between sac and duct. Twenty-three patients showed complete obstruction on syringing. Functional stenosis was detected in the other 13 cases: i.e. syringing was rendered more or less difficult, but the lacrimal passage was patent. Nevertheless, these 13 cases showed stenosis in dacryocystography. All patients had unilateral obstruction with tearing. Fifteen had a mucous discharge in the involved eye. Twenty patients had suf-

Table 6. Patient case data

Case	Sex	Age	Side	DE	PS	SL	DO	DC	MD	SN
1	F	64	L	21	Severe	LS	Functional	+		
2	F	50	R	120	Severe	LS-ND	Complete	+	+	
3	M	92	L	12	Severe	ND	Complete			Acute s.
4	F	32	L	6	Mild	ND	Functional			
5	F	75	L	157	Severe	ND	Complete	+	+	
6	F	48	L	26	Moderate	ND	Functional	+		
7	F	69	R	8	Moderate	LS-ND	Complete	+	+	
8	F	73	R	27	Severe	LS-ND	Complete	+	+	
9	F	39	R	19	Severe	ND	Complete			
10	F	57	L	9	Moderate	ND	Complete			
11	F	40	R	13	Moderate	LS-ND	Functional			
12	F	45	R	36	Severe	ND	Complete	+	+	
13	F	67	L	204	Severe	ND	Functional	+	+	Chronic s.
14	F	63	R	4	Mild	LS-ND	Functional			
15	F	58	L	13	Moderate	LS-ND	Complete	+	+	
16	M	76	R	27	Severe	LS	Complete	+	+	
17	F	16	L	20	Moderate	ND	Complete			
18	F	41	R	7	Mild	LS	Functional	+		
19	M	32	R	15	Moderate	LS	Complete			
20	M	55	R	24	Severe	ND	Complete	+	+	
21	F	60	R	120	Severe	ND	Functional	+		
22	F	66	L	240	Severe	LS	Functional	+	+	
23	F	28	L	7	Mild	ND	Functional			
24	F	61	R	240	Severe	ND	Functional	+		
25	M	36	R	49	Severe	LS-ND	Complete	+	+	
26	F	55	L	16	Severe	LS-ND	Complete	+	+	Nasal d.
27	M	65	L	14	Moderate	ND	Complete			
28	F	79	L	36	Severe	ND	Complete			
29	F	21	L	20	Severe	LS	Complete	+	+	
30	F	88	R	1	Moderate	ND	Complete			
31	F	61	L	12	Moderate	ND	Functional			Nasal d.
32	F	60	L	10	Mild	LS	Functional			
33	F	51	R	32	Severe	ND	Complete	+	+	
34	F	45	L	49	Severe	LS-ND	Complete	+	+	
35	F	70	L	29	Severe	ND	Complete			
36	F	61	L	59	Severe	LS-ND	Complete			

DC, dacryocystitis; DE, duration of epiphora (months); DO, degree of obstruction; LS, lacrimal sac; MD, mucous discharge; ND, nasolacrimal duct; PS, pathologic stage; SL, stenosis localisation; SN, sinusitis or nasal disease.

fered one to several episodes of acute dacryocystitis. Thirty-two patients were clinically without manifest nasal or sinus conditions. Two patients showed signs of nasal inflammation; one patient had acute sinusitis and one had chronic sinusitis (maxillary and ethmoidal). No significant differences were found in the degree of severity of changes versus location, severity of obstruction or presence of active inflammation. No lacrimal stones were found in any obstructed lacrimal system.

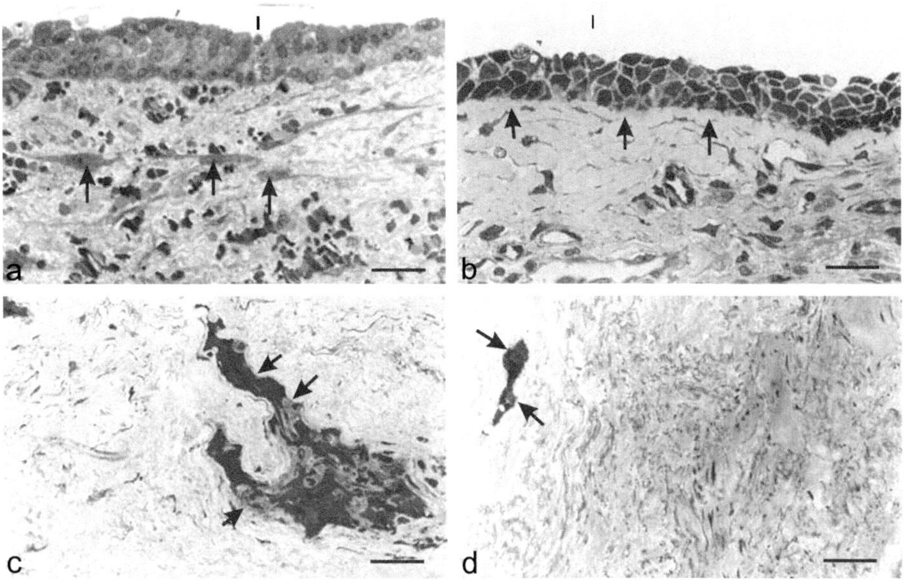

Fig. 26a–d. Histology of dacryostenosis. **a** Section through a tissue specimen of a patient showing chronic inflammation with epithelial and subepithelial infiltration of defence cells and incipient subepithelial fibrosis with increased occurrence of fibroblasts (*arrows*). *l*, Lacrimal passage; toluidine blue staining; *bar*=42 μm. **b** Section through a tissue specimen of a patient showing chronic inflammation with squamous metaplasia, absence of intraepithelial goblet cells, subepithelial fibrosis with basement membrane thickening (*arrows*), and reduction of wide capillaries. *l*, Lacrimal passage; toluidine blue staining; *bar*=42 μm. **c** Section through a subepithelial blood vessel showing chronic fibrosis with wall thickening and intimal proliferation (*arrows*). Toluidine blue staining; *bar*=44 μm. **d** Section through a subepithelial tissue showing chronic fibrosis with reduction of blood vessels (*arrows*). Toluidine blue staining; *bar*=44 μm

3.4.1 Histological Findings

Histologic examination revealed different stages of fibrosis ranging from signs of active chronic inflammation ("mild" - 5 cases) to proliferative sclerotic forms of chronic fibrosis ("moderate" - 10 cases) on up to total subepithelial fibrosis ("severe" - 21 cases).

Tissue specimens of some cases showed an active inflammatory pattern. All these patients underwent dacryocystorhinostomy (DCR) early in the course of obstruction. i.e. at 4–10 months after onset of epiphora (Table 6: cases 4, 14, 18, 23 and 32). Goblet cells and subepithelial seromucous glands demonstrated signs of hypersecretion in this early stage.

Other tissue specimens demonstrated transitional forms of chronic fibrosis with epithelial and subepithelial infiltration by defence cells that were almost lymphocytes, loss of intraepithelial goblet cells, variable loss of differentiated epithelial cells ranging from denuded epithelium to basal cell hyperplasia - often associated with squamous metaplasia - and increased subepithelial fibroblasts (Fig. 26a, b). The spe-

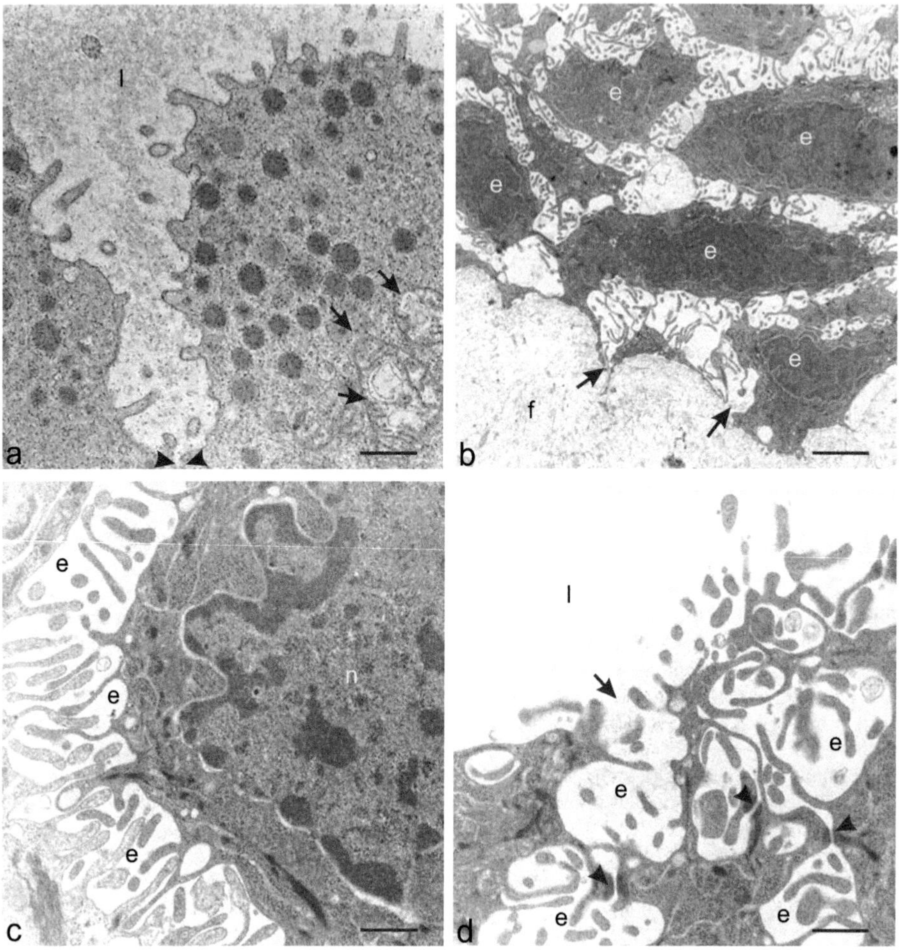

Fig. 27a–d. TEM of dacryostenosis. **a** TEM micrograph revealing an early stage of dacryostenosis. Bacteria are visible on the cell surface of two epithelial cells and in their cytoplasm infiltrating the cell organelles. *Arrows*, mitochondria; *arrowheads*, cell border; *l*, lacrimal passage; *bar*=6 µm. **b** TEM micrograph showing chronic stage of dacryostenosis. The epithelium shows squamous metaplasia. The basal metaplastic epithelial cells (*e*) are attached by cell processes to a fibrous subepithelial tissue (*f*). Spaces are visible between the cell processes (*arrows*). *Bar*=2.5 µm. **c** TEM micrograph revealing chronic stage of dacryostenosis. A metaplastic squamous cell shows a condensed, irregular nucleus (*n*), rarification of cytoplasm and reduction of cell organelles. The cell surface shows a loose microciliation projecting into a large extracellular space (*e*). *Bar*=6 µm. **d** TEM micrograph revealing chronic stage of dacryostenosis. Extracellular spaces (*e*) between metaplastic squamous cells are filled by microcilia and open directly into the lumen (*l*) of the lacrimal passage (*arrow*). The cells are attached to each other by desmosomes (*arrowheads*). *Bar*=6 µm

cialised blood vessels of the subepithelial cavernous body often showed thickening of the walls with varying degrees of intimal proliferation up to obliteration and reduction (Fig. 26c, d).

The final state of dacryostenosis was seen to be characterised by squamous metaplasia (Fig. 26b) with a total absence of goblet cells. The underlying mucosa had undergone secondary fibrosis with basement membrane thickening (Fig. 26b). Also visible was a reduction up to a total loss of the specialised blood vessels of the cavernous body (Fig. 26c, d). Tissue specimens from 16 cases with longstanding complete obstruction showed complete fibrous obliteration of the duct within the osseous canal (no lacrimal passage lumen was found, either intraoperatively or histologically).

3.4.2 Ultrastructural Findings

Early-stage dacryostenosis was characterised by active inflammation of the epithelium and subepithelial tissue (Fig. 27a). Bacteria were visible on and in epithelial cells infiltrating all intracellular organelles (Fig. 27a). Nevertheless, in addition to a soft oedema in this stage, epithelial cells revealed a nearly normal aspect with double-layered epithelium and microciliation of the upper cells (Fig. 27a). Goblet cells and subepithelial seromucous glands found in some specimens revealed signs of hypersecretion. The lamina propria was seen to be oedematous and the cavernous body more prominent with reactive hyperaemia.

In the chronic state, all specimens showed variable loss of differentiated epithelial cells ranging from denuded epithelium to basal cell hyperplasia - often associated with squamous metaplasia. These cells were of irregular shape, in some cases showing long cell processes with condensed, irregular nuclei (Fig. 27b). Also, the cytoplasm was very dense with rarification of mitochondria and no visibility of endoplasmic reticulum or Golgi apparatus (Fig. 27c). The whole surface showed a loose microciliation with large transcellular spaces (Fig. 27b–d) measuring between 5 up to 25 μm (possibly due to tissue shrinkage in fixation). These spaces opened without any border into the lumen of the lacrimal passage (Fig. 27d) and led basally to an oedematous-fibrotic lamina propria (Fig. 27b). Cells were connected to each other by cell processes, at which point desmosomes were visible (Fig. 27d). Besides this cell type, several lymphocytes showing signs of degeneration were visible in the epithelium.

Conclusions (3.4). Descending inflammation from the eye or ascending inflammation from the nose initiates swelling of the mucous membrane, remodelling of the helical arrangement of connective tissue fibres, malfunctions in the subepithelial cavernous body with reactive hyperaemia, and temporary occlusion of the lacrimal passage. In the follow-up, repeated isolated occurrence of dacryocystitis leads to structural epithelial and subepithelial changes, which may lead either to a total fibrous closure of the lumen of the efferent tear duct or to a non-functional segment in the lacrimal passage that is manifest on syringing.

4 Discussion

4.1
Mechanisms of Immune Defence in the Nasolacrimal Ducts

4.1.1
The Function of the Epithelium Lining the Nasolacrimal Ducts

4.1.1.1
Lectin Binding Sites and Mucins

In the epithelium of the lacrimal sac and nasolacrimal duct, goblet cells are present (Werncke 1905), which sometimes form mucous glands (Paulsen et al. 1998; Thale et al. 2001). The secretory product of the mucus layer formed by these cells has been attributed largely to immunological response (Paulsen et al. 1998; Thale et al. 2001). It contains carbohydrates including galactose, *N*-acetyl-glucosamine, fucose and *O*-acetylated as well as non-*O*-acetylated sialic acids (Table 2, Fig. 2i, j). Sialic acids are present in α-(2–6)-linkage in both goblet and epithelial cells, whereas α-(2–3)-linkages are detectable only in epithelial cells (Paulsen et al. 1998; Table 2, Fig. 2i, j).

The high occurrence of mucin-producing goblet cells is of particular interest. It has been suggested that mucus serves several functions, including lubrication of the mucosa, water-proofing to regulate epithelial cell hydration and protection of mucosal surfaces against potentially harmful substances such as particles, aggressive chemical agents, digestive enzymes, food lectins, toxins, and bacterial and other infectious agents (Faillard and Schauer 1972; Hutch 1970; Walker 1976; Forstner 1978; Schauer 1992; Reuter et al. 1992). That mucus protects mucosal surfaces from infective agents and noxious substances has long been assumed. Intestinal mucus has been observed to carry away bacteria (Florey 1933). Moreover, it has been shown that mucus possesses structures mimicking the receptor sites for micro-organisms on epithelial cells that facilitate trapping and subsequent disposal of bacteria (Abraham and Beachey 1985) and viruses (Reuter et al. 1988). A variety of oral and intestinal bacteria has been revealed to produce neuraminidase (sialidase), an enzyme that can degrade mucins by the removal of sialic acid (Corfield 1992). In addition, oral and intestinal bacteria synthesise an array of other glycosidases that can attack the oligosaccharide residues of mucins (Schauer 1997).

Higher affinity interactions of "lectin-like" receptors on the micro-organism with ligands of epithelial cell surfaces are, however, essential for colonisation (Gibbons 1977; Gibbons and van Houte 1980). Recently, it has been shown that bacterial receptors are involved in signal transduction at the molecular level or in mimicking the

Table 7. Production of mucins (MUC) in human nasolacrimal ducts

Abbreviation	Epithelial cells	Goblet cells	Specials
MUC1	No	No	Positive results in RT-PCR analysis
MUC2	Yes	Yes	Only in small cell-conglomerates
MUC3	(Unknown)	(Unknown)	
MUC4	No	No	Positive results in RT-PCR analysis
MUC5AC	No	Yes	
MUC5B	No	Yes	
MUC6	No	No	
MUC7	Yes	No	
MUC8	(Unknown)	(Unknown)	

Results are based on unpublished immunohistochemical results.

function of growth factor ligands either coupled to tyrosine kinase activity or to heterotrimeric G protein. The factors lead to a multiplicity of cellular events following their activation depending on factors such as cellular type, species and/or tissue (Hebert 2000). The bacterial receptors may be sugar-binding proteins on microvilli, or surface-bound enzymes with specific bindings properties. They have been found to bind to glycoprotein or glycolipid ligands on mucosal cell surfaces (Gibbons 1977; Jones 1977; Kelm and Schauer 1997; Gibbons and von Houte 1980). For example, it has been observed that for blocking rotavirus infection of MA104 cells, galactose-specific lectins were the most inhibitory (Jolly et al. 2000). Moreover, in vitro experiments give evidence that interaction of Sambucus nigra (SNA) with human neutrophils is accompanied by dose-dependent release of lysozyme (Gorudko and Timoshenko 2000). Lectins on macrophages can contribute to host defence by means of mannose receptor (MR) and complement receptor three (CR3), both of which are able to mediate phagocytosis of pathogenic microbes and induce intracellular killing mechanisms (Linehan et al. 2000).

The tear film overlying the ocular surface is composed of three components: an outer lipid component secreted by the meibomian glands; an aqueous component secreted by the lacrimal gland and the accessory lacrimal glands; and an mucus component containing mucins as its major structural ingredient (Berry et al. 1996). These mucins are thought to influence the rheological properties of the ocular mucus (Corfield et al 1997; Gipson and Inatomi 1997). The rheological properties are defined in terms of tear break-up time, which is changed in various pathological conditions (e.g. in patients with dry eye symptoms). Alterations of mucin in the human conjunctival epithelia of such patients have been reported (Danjo et al. 1998).

So far, unpublished observations show that the mucin composition of the nasolacrimal ducts differs from mucins of the ocular surface. Thus, our group recently could show in collaboration with the Mucin Research Group of Bristol, UK, that mucins MUC2, MUC5AC, MUC5B, and MUC7 are actually produced in the human nasolacrimal ducts (Table 7).

It has also been suggested that IgA supplements the protective activity of the mucus by incorporation into the mucus layer of mucosal surfaces (Bienenstock 1974). It must be mentioned here that the molecular nature of mucin–IgA interactions is poorly understood, and binding has been difficult to demonstrate experimentally with purified components (Kilian et al. 1988). IgA of various molecular properties

Table 8. Predominant TFF-peptide/secretory mucin combinations observed in various human mucin-producing cells

Cell	TFF peptide/mucin
Conjunctival goblet cell	TFF1+3/MUC5AC
Submandibular gland: serous cell	TFF3/MUC7
Respiratory goblet cell	TFF3/MUC5AC
Respiratory submucosal glands	TFF3/MUC5B
Gastric surface cell	TFF1/MUC5AC
Gastric mucous neck cell	TFF2/MUC6
Intestinal goblet cell	TFF3/MUC2
Nasolacrimal ducts: goblet cell	TFF1/MUC5AC+MUC5B+MUC2
Nasolacrimal ducts: columnar epithelial cell	TFF3/MUC7+(MUC2)
Nasolacrimal ducts: serous gland cell	TFF3/MUC7

and origins can interact with functionally diverse cells, including epithelial cells, B and T lymphocytes, NK cells, cells of the monocyte/macrophage lineage and neutrophils (Mestecky and McGhee 1987).

4.1.1.2
TFF Peptides

Recently, it has been shown that goblet cells of porcine and human conjunctiva secrete TFF peptides (Jagla et al. 1999; Langer et al. 1999, 2001; 2002) (formerly P-domain peptides, trefoil factors) (Wright et al. 1997) which are, together with mucins, typical constituents of mucus gels (e.g. from the gastrointestinal and the respiratory tracts and the uterus; for reviews, see Hoffmann et al. 1993, 2001a,b). Three TFF peptides have been characterised in mammals, including human beings: TFF1 (formerly pS2), TFF2 (formerly hSP), and TFF3 (formerly hP1.B/hITF). They are characterised by the TFF motif, a three-looped structure held tightly together by disulphide bonds based on six cysteine residues. One such motif is found in TFF1 and TFF3, while TFF2 possesses two TFF domains. Besides their occurrence as major secretory products of many mucin-producing cells (Hoffmann et al. 2001; Hoffmann and Jagla 2001) they have also been detected in the brain (Jagla et al. 2000).

The physiological functions of TFF peptides are multiple (Hoffmann et al. 2001b). They promote migration of intestinal, corneal or bronchial epithelial cells in vitro (Dignass et al. 1994; Göke et al. 2001; Oertel et al. 2001), they have anti-apoptotic properties (Chen et al. 2000; Kinoshita et al. 2000; Taupin et al. 2000), and they induce cell scattering (Williams et al. 1996; Emami et al. 2001). TFF3 has also been detected as a new neuropeptide of the hypothalamo-pituitary axis (Jagla et al. 2000; Schwarz et al. 2001). All TFF peptides are known for their protective or healing effects in vivo, in particular for the gastrointestinal mucosa, and they are aberrantly secreted during various chronic inflammatory diseases (for reviews, see Wright et al. 1998; Hoffmann et al. 2001; Hoffmann and Jagla 2001).

In the present study, the biosynthesis of TFF peptides on the mucosal side of the nasolacrimal ducts was analysed. Biopsy material as well as tissue from cadavers was studied and compared. The data suggest that the human nasolacrimal ducts synthesise and store the secretory peptides TFF1 and TFF3, but not TFF2.

TFF3 is actually produced by the columnar epithelial cells of the nasolacrimal ducts as well as the serous cells of seromucous glands, but is absent in goblet cells. This result is in agreement with a similar observation concerning TFF3 in the major salivary glands (Jagla et al. 1999a), especially in submandibular and sublingual glands, but is surprising in so far as the distribution pattern is in clear contrast to the human and porcine conjunctiva (Langer et al. 1999, 2001, 2002), in which the peptide is typically localised in goblet cells. Furthermore, in most other epithelia, e.g. of the gastrointestinal and respiratory tract (Hoffmann et al. 2001; Hoffmann and Jagla 2001), TFF3 is normally found in goblet cells (Table 8).

Interestingly, several publications describe occasional supranuclear TFF1 or TFF3 immunoreactivity (Robbins et al. 1993; Taupin et al. 1996, 2001). Also, the present immunohistochemical results reveal nuclear TFF3 staining in single sections (Fig. 6g), which was not inhibited with a corresponding synthetic peptide (Fig. 6h), whereas all cytoplasmic staining was inhibited. This points to the presumption that the nuclear staining is due to an unspecific reaction and would therefore seem to be artificial. By contrast, the present immunohistochemical results also reveal TFF1 staining in direct proximity to the nucleus in several sections where the peptide was detected (Fig. 6a). However, this staining was inhibited with a corresponding synthetic peptide (Fig. 6b) and would therefore appear to be specific.

The distribution pattern of TFF3 in the nasolacrimal ducts does not overlap with that of TFF1. This finding is an agreement with most other mucous epithelia (Table 8). TFF1 of the nasolacrimal ducts is localised in goblet cells. This situation can be compared with that in the conjunctiva (Langer et al. 1999, 2001, 2002). By contrast, TFF1 in the stomach is typically present in gastric surface cells (Rio et al. 1988; Levebvre et al. 1993; Jagla et al. 1999a). The finding that TFF1 is present only in trace amounts in human nasolacrimal ducts is reminiscent of the situation in salivary glands (Jagla et al. 1999a) and to descriptions in previous reports (Rio et al. 1988; Pigott et al. 1991).

Each mucin-producing cell type has been shown to secrete a characteristic TFF-peptide/mucin combination (Hoffmann et al. 2001; Hoffmann and Jagla 2001), probably reflecting the complex physiological needs of the environment of these cells (Table 8). The localisation of TFF1 and TFF3 in the human conjunctival goblet cells matches precisely that of the secretory mucin MUC5AC (Inatomi et al. 1996; Gipson and Inatomi 1997; Jumblatt et al. 1999). The localisation of TFF3 in serous cells of the submandibular glands is coincidental with that of mucin MUC7 (Nielsen et al. 1996, 1997). Thus, the presence of the mucins MUC5AC and MUC7 in the nasolacrimal ducts is within the limits of expectation and seems to be true (Tables 7 and 8).

Based on the co-secretion of TFF peptides with mucins, it is postulated that TFF peptides interact with mucins as "link peptides", influencing the rheological properties of these complex viscous biopolymers (Hauser et al. 1993). Preliminary studies with TFF2 and TFF3 underline this hypothesis, as both increase the viscosities of mucin preparations (Babyatsky et al. 1996). The dimeric structure of TFF1 (Chadwick et al. 1997) and TFF3 (Chinery et al. 1995) would be ideally suited to form an entangled network (Verdugo 1990; Hoffmann and Joba 1995) with the mucins of the nasolacrimal ducts. The precise nature of the interaction between TFF peptides and mucins is currently not known; however, recent studies showed that TFF1 interacts with the von Willebrand Factor (VWBF) C-domains of MUC2 and MUC5AC indicating that the protective effect of TFF peptides may operate by organising the complex mucus layer (Tomasetto et al. 2000).

The differences in TFF-peptide distribution between conjunctiva and nasolacrimal ducts are interesting with regard to tear transport through the lacrimal passage. The change in the composition of the mucus layer inside the nasolacrimal ducts may have a significant impact on the rheological properties affecting the drainage of tears through the lacrimal passage.

TFF1 and TFF3 in the nasolacrimal ducts are generally expected to have protective effects as in the gastrointestinal tract (for review, see Hoffmann et al. 2001; Hoffmann and Jagla 2001). It is interesting in this regard that the columnar epithelial cells of the nasolacrimal ducts also produce a broad spectrum of antimicrobial peptides including lysozyme, lactoferrin, secretory phospholipase A_2, human β-defensin 1 and human-inducible β-defensin 2 (see Sect. 4.1.1.3; Paulsen et al. 2001a, 2002c). An interaction of TFF3 with the antimicrobial peptides is possible as both are products of the same cells. Remarkably, a combined secretion of TFF peptides and lysozyme is also observed in a specific gland-like structure termed the ulcer-associated cell line (UACL) (Wright 1998). The anti-apoptotic effect of TFF3 (Chen et al. 2000; Taupin et al. 2000) should also be considered in this context.

Cases of functional dacryostenosis, i.e. patients with epiphora in spite of patent lacrimal passages on syringing, may result not only from the occurrence of a non-functional segment in the lacrimal passage (Paulsen et al. 2001b), but also from the downregulation of TFF peptides and mucins in this segment. Moreover, changes in TFF-peptide production may contribute to the occurrence of dacryoliths in the lacrimal passage. TFF-peptide secretion might also be influenced by alterations in glycosylation of goblet cell and epithelial cell mucins, since they occur in patients with dry eye symptoms (Danjo et al. 1998). However, no data are available as yet concerning the synthesis and secretion of ocular TFF peptides during pathological conditions.

4.1.1.3
Antimicrobial Peptides

Dacryocystitis is the most frequent disease of the efferent lacrimal system. Its course may be either acute or chronic and slow to resolve even with systemic antibiotic therapy (Cahill and Burns 1993). Like the cutis, all mucosal surfaces of the body are in constant interaction with environmental micro-organisms and hence vulnerable to infection. The human mucosa has developed a variety of anti-infection strategies to prevent micro-organisms from entering through surfaces and to thwart attacks from micro-organisms that have already entered the epithelia.

Recent evidence indicates that the surface of the efferent tear ducts is an integral part of the mucosal immune system (Perra et al. 1995; Paulsen et al. 1998, 2000b, 2002b; Knop and Knop 2000). The protection of the lacrimal passage from microbial attack must therefore have been of paramount importance for the survival of the individual and the preservation of the species.

During recent years, endogenous antimicrobial peptides, active against gram-positive and gram-negative bacteria, have attracted much interest, in the ocular system as well (Cullor et al. 1990; Schwab et al. 1992; Gottsch et al. 1998; Haynes et al. 1998, 1999, 2000; McNamara et al. 1999; Lehmann et al. 2000; Zasloff 2002). Moreover, some studies suggest that invertebrate epithelial cells can recognise micro-organisms and mount a rapid defence response through the production of various inducible antibiotic peptides, as shown for Drosophila (Tzou et al. 2000). This leads to a characteristic broad spectrum of antimicrobial activity against bacteria and fungi (Zasloff 2002).

In the present study, we sought to identify and characterise antimicrobial peptides released by the human tear duct mucosal cells in healthy state and to investigate potential differences in the presence of dacryocystitis. The human nasolacrimal ducts produce a spectrum of antimicrobial peptides including the human inducible defensin HBD-2, which is produced after treatment of epithelial cells with proinflammatory cytokines or contact with pathogenic bacteria such as the mucoid form of *Pseudomonas aeruginosa* (Harder et al. 1997; Schröder 1999; Schröder and Harder 1999; Harder et al. 2000).

It appears that lysozyme, lactoferrin and sPA_2 are actually produced by the mucosal surface of the lacrimal sac and nasolacrimal duct as well as the serous cells of seromucous glands. Lysozyme and lactoferrin are well known as constituents of tear protein (Janssen and van Bijsterveld 1983). Lysozyme is a low-molecular-weight protein showing bacteriostatic and bactericidal activity. It enzymatically cleaves the glycosidic bond of muramic acid residues in the cell walls of certain bacteria. Lysozyme is especially effective in cytolysis of gram-positive organisms. Moreover, it enhances the antibacterial action of complement and T cells against gram-negative bacteria (Wilhelmus 1985). Lactoferrin is an iron-binding protein that reduces the amount of free iron available to bacteria. It provides both bacteriostatic and bactericidal protection (Oram and Reiter 1979). Furthermore, it has been shown to play a role in primary antibody response, lymphocyte proliferation, cytokine production, NK cell activity and the regulation of complement activation (Kijlstra 1990). The calcium-dependent enzyme secretory PLA_2 has recently been revealed as an important tear component of host defence systems against many gram-positive pathogens (Qu and Lehrer 1998).

This epithelial antimicrobial defence is supported by neutrophils present in large amounts inside the epithelium and the subepithelial connective tissue of the lacrimal sac and nasolacrimal duct (Paulsen et al. 1998, 2000b; Knop and Knop 2000, 2001). It is speculated that this abundant occurrence of neutrophils could be based on the fact that the efferent tear ducts have to drain all antigens taken up by the ocular surface (Knop and Knop 2000). The neutrophils are positive for α-defensins 1, 2 and 3, which have been shown to be amply present in these cells (Ganz et al. 1985; Ganz and Lehrer 1995; Hill et al. 1991). α-Defensins 5 and 6 were not detected in the lacrimal passage. Therefore, on the information currently available, α-defensins 5 and 6 remain the products of small intestine Paneth cells only.

In addition to α-defensins 1, 2, and 3, PCR analysis revealed BPI and CAP37 to be expressed in probes of the nasolacrimal epithelium. It seems that BPI and CAP37 are also released into the ocular surface fluids by resident or passing neutrophils, since BPI – a 55-kDa basic protein found in azurophilic granules of polymorphonuclear leukocytes (Marra et al. 1992) and, more recently, on the surface of neutrophils (Weersink et al. 1993), in eosinophils (Calafat et al. 1998), in plasma (White et al. 1994) and in inflammatory fluids – unfolds a highly cytotoxic action against many gram-negative bacteria. CAP37, a human polymorphonuclear leukocyte granule-derived 37-kDa protein, has also been identified as possessing antimicrobial activity against gram-negative bacteria (Shafer et al. 1986). In addition, there are reports identifying antimicrobial peptides such as α-defensins 1 and 2 and CAP37 as potent chemoattractants for T cells (Hancock and Lehrer 1998).

Apart from α-defensins, mammals produce a second family of defensins, which due to their structural similarity to α-defensins are termed the β-defensin family. β-defensins, which occur as ~4-kDa peptides containing 38–42 amino acids, are highly cationic, variably arginine-rich and are distributed in a greater variety of epithelia than α-defensins (Lehrer et al. 1991; Diamond and Bevins 1998). The two β-defensins analysed in the present study, HBD-1 and HBD-2, are thought to exert their antimicrobial activity by interacting with membranes of metabolically active bacteria, perhaps by forming pores and causing membrane disruption (Lehrer et al. 1989; Hill et al. 1991; Kagan et al. 1994; Ganz and Lehrer 1995). Other possible roles could include promotion of nasolacrimal duct epithelial healing (Murphy et al. 1993), monocytic (Territo et al. 1989), dendritic and T-cell chemotaxis (Yang et al. 1999; Chertov et al. 2000; Yang et al. 2000), synergistic activity with lysozyme and lactoferrin (Bals et al. 1998), and complement activation (Prohaszka et al. 1997; Prohaszka et al. 1998). Comparable to other mucosal sites (Bals et al. 1998; Singh et al. 1998; Zucht et al. 1998; Haynes et al. 2000), our PCR results reveal that HBD-1 is expressed in all healthy lacrimal systems, although we were unable to detect the peptide in all healthy samples analysed by immunohistochemistry, perhaps because its concentration is below the limit of detection of our antibody. In the presence of inflammation, HBD-1 also appears to be expressed in the lacrimal system. However, as shown in a recent study, early infections can reduce or turn off expression of antimicrobial peptides (Islam et al. 2001) suggesting that its production depends on the status of the local bacterial microflora. Our results underline this finding, since HBD-1 and CAP37 were sometimes not detected in acute dacryocystitis in particular (Table 3).

HBD-2, one of the only two known human inducible defensins, which is upregulated by contact with gram-negative and gram-positive bacteria as well as *Candida albicans* (Harder et al. 1997; Harder et al. 2000) or generated in response to inflammatory cytokines during infection (Haynes et al. 2000), is only detected in the lacrimal passage in the presence of bacterial dacryocystitis. The inducibility of HBD-2, and the fact that it is approximately 10 times as potent as HBD-1 (Singh et al. 1998) and shows a wider antibacterial spectrum, makes HBD-2 a stronger candidate for antimicrobial defence in the efferent tear ducts, despite the salt content of tears which approaches that of serum (Milder 1978). Singh et al. (1998) demonstrated that the antimicrobial activities of both HBD-1 and HBD-2 were reduced by NaCl. Thus, cytokine-induced HBD-2 production may be more important for antimicrobial defence than constitutive HBD-1 production.

Besides the surgically obtained samples from patients with dacryocystitis, one sample from a body donor (Table 4) also revealed a PCR product for HBD-2. It is not known whether this was due to technical factors or more likely reflects a bacterial, viral, fungal, allergic or sterile infection during the last days of the person's life. Taken together, the present study suggests that both HBD-1 and HBD-2 play important, although somewhat different, roles in the mucosal defence of the efferent tear ducts.

No expression was found for the peptide LL-37. It is not clear whether this was due to technical factors or reflects the absence of this inducible peptide in the nasolacrimal ducts. The latter is more likely as there are no current publications available on LL-37 in ocular tissue.

Our data suggest that the human nasolacrimal ducts produce a spectrum of antimicrobial peptides. Such production of several antimicrobial peptides by different cells is more effective in antimicrobial defence, as the combined action of antimicrobial peptides leads to synergistic and additive killing effects (Singh et al. 2000). An understanding of the exact mechanism of production and regulation of antimicrobial peptides at the mucosal surface of the lacrimal passage will provide further insight into the occurrence of dacryocystitis, which often leads to residual functional impairment with epiphora. The factors controlling the production of nasolacrimal duct-associated antimicrobial peptides are unknown and it is likely that some infection risk factors like old age, changes in hormonal status (post-menopausal women), a narrow bony channel, or immunodeficiency are associated with downregulation of antimicrobial peptide production. Another hypothesis suggests that a pre-existing stenosis or narrowing of the lacrimal passage downregulates the production of antimicrobial substances: the normally constant flow of tears could be a positive feed-back signal for production, which comes to a halt if tears are not drained into the nose. This does not, off course, explain why some patients with epiphora due to postsaccal stenosis never develop dacryocystitis.

Moreover, it may be hypothesised that purified or recombinant antimicrobial peptides, especially inducible, may be ideal agents in the therapy of dacryocystitis when applied topically via injection through the lacrimal canaliculi directly to the site of infection. Nevertheless, there is a lack of experience in the clinical use of cationic peptides and this important aspect should be addressed in future investigations, especially because a few reports reveal that defensins could accelerate two key events involved in scarring processes of the lacrimal drainage passages such as dacryostenosis, i.e. by promoting cellular proliferation (Kudryashov et al. 1990; Murphy et al. 1993) and fibrin formation (Higazi et al. 1995, 1996).

4.1.2 Specific Immune Defence by MALT

4.1.2.1 MALT of the Nasolacrimal Ducts

The term MALT is most often associated with the well-defined lymphoid structures in the gut wall, the Peyer's patches. According to Isaacson (1992), MALT is typified by Peyer's patches in the terminal ileum and comprises four lymphoid compartments as follows: (1) organised mucosal lymphoid tissue, (2) a lamina propria, (3) intra-epithelial lymphocytes and (4) the mesenteric lymph nodes. The organised lymphoid tissue comprises a B- and T-cell component. The B-cell component consists of a follicle with a centre and mantle zone and a more or less well-developed marginal zone. There is also an interfollicular area of T cells with high endothelial venules. The lamina propria is diffusely infiltrated by plasma cells, most of which are synthesising IgA, and lymphocytes, most of which are $CD4^+$ T cells with some memory B cells, macrophages and other accessory cells (Isaacson 1992).

In humans, as already described, the lacrimal sac and the nasolacrimal duct contain a double-layered epithelium resting on a broad basement membrane. Perra et al. (1995) demonstrated large amounts of IgA inside and on the surface of the lining epithelium of the nasolacrimal ducts which could interact with T and B lymphocytes and macrophages amply present inside the epithelium and the underlying substantia propria (Markovich 1951; Tsuda 1952; Paulsen et al. 1998; Knop and Knop 2001; Schaudig et al. 2001; Thale et al. 2001). Duke-Elder (1961) described a layer of adenoid tissue beneath the epithelium, sometimes aggregated into follicles. The adenoid tissue has been shown to consist of immunocompetent cells differentiated as T and B lymphocytes and macrophages showing a special distribution inside the lamina propria (Paulsen et al. 1998).

In the human nasolacrimal ducts, we were able to demonstrate subepithelial visibility of aggregated follicles in some of the efferent tear ducts investigated (Figs. 9 and 12; Paulsen et al. 2000b, 2002b). These aggregations fulfilled the criteria for designation as MALT. They consist of organised mucosal lymphoid tissue characterised by the presence of reactive germinal centres and mantle zones. Around the mantle zone, one can identify an additional zone of somewhat larger cells corresponding to marginal zone cells. These cells extend into the overlying epithelium, forming a lymphoepithelium (Paulsen et al. 2000b, 2002b). According to the terminology of MALT in other body regions, MALT of the human nasolacrimal ducts was termed LDALT (lacrimal drainage-associated lymphoid tissue) by Knop and Knop (2001). We have termed it TALT (tear duct-associated lymphoid tissue; Paulsen et al. 2000b, 2002b,d).

Specific secretory immunity depends on sophisticated co-operation between the mucosal B-cell system and an epithelial glycoprotein called the secretory component (Brandtzaeg 1992). Initial stimulation of Ig-producing B cells is believed to take place mainly in organised MALT (Butcher and Picker 1996). It has become evident that considerable regionalisation or compartmentalisation exists in MALT, perhaps being determined by different cellular expression profiles of adhesion molecules and/or the local antigenic repertoire. Antigenic stimulation of B cells results in the generation of predominantly IgA-synthesising blasts that leave the mucosae via efferent lymphatics, pass through the associated lymph nodes into the thoracic duct and enter the circulation. The cells then return selectively to the lamina propria as plasma cells or memory B cells (Isaacson 1992) by means of homing mechanisms.

The lamina propria within the nasolacrimal ducts is normally devoid of organised lymphoid structures, suggesting that its immunological effector functions depend on the generation of immune responses elsewhere. In this context, experimental work has focused on the paired lymphocytic cell aggregates present at the entrance to the nasopharyngeal duct in rodents (Kuper et al. 1992). This organised lymphatic tissue has many similarities to gut-type MALT, and its inductive function may be crucial to the immune response of mucosal surfaces in the head and perhaps beyond this region.

In our studies, organised MALT was found in nearly a third of nasolacrimal ducts from unselected body donors with unknown previous history of disease regarding the eye, efferent lacrimal pathway or nose (Fig. 10)(Paulsen et al. 2000b, 2002b). This suggests that MALT is not a ubiquitous finding in the efferent lacrimal system and is rather acquired in response to antigenic stimulation. It may develop in response to bacterial or viral infections or as a result of allergic reactions. The development or acquisition of MALT in individual human efferent tear ducts remains unclarified, but, once present, it can provide the basis from which primary low-grade B-cell lymphoma of the MALT type may arise (Paulsen et al. 2000b, 2002b) as they have been shown by Kheterpal et al. (1994), White et al. (1995), and Tucker et al. (1997).

4.1.2.2 Loss of TALT in Association with the Scarring of Symptomatic Dacryostenosis

It has been shown above that the mucosa of the nasolacrimal ducts plays a significant role in unspecific immune defence by producing a mucus layer containing different carbohydrates (Paulsen et al. 1998; Thale et al. 2001), TFF peptides (Paulsen et al. 2002a) and antimicrobial peptides (Paulsen 2001a, 2002c). As demonstrated above, the surface of the efferent tear ducts is also an integral part of the specific mucosal immune system and belongs to the mucosa-associated lymphoid tissue (MALT) (Paulsen et al. 2000b, 2002b).

However, it is yet not clear whether MALT of the nasolacrimal ducts is a normal component of the human efferent tear ducts. As already discussed, MALT of the efferent tear ducts may develop in response to bacterial or viral infections or as a result of allergic reactions (Paulsen et al. 2000b). Moreover, the description of primary low-grade B-cell lymphomas in the lacrimal passage makes this question clinically relevant (Kheterpal et al. 1994; White et al. 1995; Tucker et al. 1997).

To test the hypothesis that MALT of the nasolacrimal ducts is acquired in response to bacterial stimulation, nasolacrimal ducts from cadavers with unknown prior history of efferent tear duct, ocular and nasal disease were examined along with biopsies from patients suffering from post-inflammatory dacryostenosis and compared with a view to presence of lymphoid tissue with the structure and phenotype of organised MALT.

The findings reveal MALT in 34% of "healthy" cadavers investigated. The TALT comprises a B- and T-cell component. The B-cell component consists of a follicle with a centre and mantle zone and a more or less well-developed marginal zone. There is also an interfollicular area of T cells with high endothelial venules. The lamina propria is diffusely infiltrated by plasma cells, most of which are synthesising IgA, and lymphocytes, most of which are T cells with some B cells, macrophages and other accessory cells.

In comparison with MALT of the gastrointestinal tract (GALT), it is as yet not known whether M cells occur in TALT (M cells are highly specialised epithelial cells that facilitate uptake of macromolecular substances from the gut. They are able to present to lymphocytes and macrophages specific, potentially pathogenic epitopes of the macromolecules. The immune system needs this direct contact with the epitopes of the macromolecules to initiate a specific immune response) (Gebert et al. 1996). Also, the functional meaning of the structures is not known since, in contrast to Peyer's patches, TALT is only found in 30%–40% of "healthy adult specimens" (Paulsen et al. 2000b, 2002b; Knop and Knop 2001).

As early as 1988, Hyjek et al. (1988) postulated that organised lymphoid tissue from outside the gut was supposed to form MALT in response to antigenic stimulation, which may contribute to an autoimmune condition such as Sjögren's syndrome in the salivary gland. Moreover, it was demonstrated that, in the conjunctiva

(Wotherspoon et al. 1994), nose (Kuper et al. 1992), larynx (Tschernig et al. 1995) and lung (Emery and Dinsdale 1973; Pabst and Gehrke 1990; Gould and Isaacson 1993), MALT is absent in neonates and is acquired in early childhood in response to antigenic stimulation. Animal experiments indicate that the development of gastric (Wotherspoon 1991), bronchus (Delventhal et al. 1992a,b), or nasopharyngeal (Kuper et al. 1992) -associated lymphoid tissues is acquired specifically in response to local infection. By contrast, a retrospective study in specimens removed during surgical procedures on patients with chronic pulmonary inflammation revealed bronchus-associated lymphoid tissue (BALT) to be generally less prominent than that in animals (Delventhal et al. 1992) and Wotherspoon et al. (1994) advise caution when extrapolating animal experimental data to humans, since the distribution and structures of MALT vary among different species.

The relevance of our present findings is limited by the size of the specimen samples investigated. We cannot exclude the possibility that cases classified as "without TALT" did contain TALT in other parts of the lacrimal system that were not investigated. Moreover, it is not possible to determine decisively whether an inflammation of the lacrimal passage has occurred in specimens from cadavers. However, the congruency between the size of specimens from patients and cadavers and the numbers of lacrimal systems from cadavers investigated partly compensate these disadvantages. Thus, the present results lead to the assumption that chronic inflammation of the mucosa and subepithelial mucous membrane is probably not an essential factor in the development of TALT and therefore does not necessarily lead to its formation, suggesting that its immunologic effector functions depend on the generation of immune response elsewhere.

At any rate, a recently published paper has certainly put our understanding of MALT in a different light concerning the functional significance of TALT. Alpan et al. (2001) demonstrated that a systemic immune response to orally administered soluble antigens does not depend on the presence of functional GALT, but more likely on initiation of immune response by gut-conditioned dendritic cells.

Based on the most recent description of TALT, it has yet not been taken into consideration that antigens drained by the tear fluid themselves might be able to induce a kind of immune deviation. This would be plausible with regard to protection of the cornea against inflammatory destruction analogous to what is observed in the nervous system (Wenkel et al. 2000) and the anterior eye chamber (Streilein and Niederkorn 1985).

On the other hand, the drainage pathways of antigens from the anterior eye chamber, the location of their presentation and the passage of the corresponding antigen-presenting cells are unknown at present. Especially the role of the TALT (Paulsen et al. 2000b, 2002b; Knop and Knop 2001) and also of the conjunctiva-associated lymphoid tissue (CALT) (Österlind 1944; Axelrod and Chandler 1979; Wotherspoon et al. 1994; Dua et al. 1994; Chodosh et al. 1998; Knop and Knop 2000), which together have been designated "eye-associated lymphoid tissue" (Knop and Knop 2001), for the immune privilege of the anterior eye chamber, remains unclear.

It can be concluded that the development of MALT is a common feature that is often found in symptomatically normal nasolacrimal ducts. As TALT seems to be lost associated with the scarring of symptomatic dacryostenosis, it is unlikely that the presence per se of TALT leads to scarring. Whether special types of bacteria, viruses or other factors such as some kind of immune deviation are responsible for the development of TALT in humans needs to be investigated in future prospective and experimental studies.

4.2 Mechanisms Regulating Tear Flow Through the Nasolacrimal Ducts

4.2.1 Regulation of Tear Flow by the Cavernous Body

The lumen of the lacrimal sac and the nasolacrimal duct is surrounded by a vascular plexus (Henle 1866). This network of wide-lumened blood vessels that has not been investigated until now is connected caudally with the cavernous body of the nasal inferior turbinate (Thale et al. 1997). The surrounding vascular plexus contains specialised blood vessels and is comparable to a cavernous body (Paulsen et al. 2000d).

The blood vessels are specialised arteries, venous lacunae, and veins. The arteries contain two muscle layers. They are known from skin, oesophagus, and ovary (Stieve 1928) as well as the lower anterior spinal artery (Parke et al. 1995) and are called barrier arteries, in accordance with the German term *Sperrarterien*. Their function is to reduce or interrupt the blood supply to the downstream blood vessels. The arteries split just beneath the epithelium into superficial arcading branches. A dense network of capillaries arises from these branches to supply blood to seromucous glands of the lamina propria and also to bring nutritive substances to the epithelium. The blood from the capillary network is collected by short postcapillary venules that drain into widely convoluted venous lacunae. These blood vessels are called capacitance veins (Temesrékàsi 1969) because of their probable ability to store large amounts of blood (Paulsen et al. 2000d). Segments of the capacitance veins are sometimes narrowed. The tunica media of these segments contains a muscle layer of helically arranged smooth muscle cells that effects closure of the segment. In agreement with the nomenclature of the nose (Körner 1937) these appliances are called cushion veins, which are also termed throttle veins in the English literature and are referred to as *Polstervenen*, *Sperrvenen*, or *Drosselvenen* in the German literature. They can reduce or interrupt venous blood outflow and allow large amounts of blood to accumulate inside the capacitance veins. Finally, blood is collected by large veins that drain the blood out of the lacrimal passage. Furthermore, arteriovenous anastomoses are seen to connect branches of the arteries with capacitance veins (Paulsen et al. 2000d).

The specialised blood vessels may facilitate closure and opening of the lumen of the lacrimal passage by swelling and shrinkage of the cavernous body. Swelling occurs when the barrier arteries are opened and the throttle veins closed. Filling of the capacitance veins occurs at the same time with closure of the lumen of the lacrimal passage. By contrast, closure of the barrier arteries and opening of the throttle veins reduces the blood stream to the capacitance veins, simultaneously allowing blood outflow from these veins with resultant shrinkage of the cavernous body and dilatation of the lumen of the lacrimal passage. Arteriovenous anastomoses provide for direct blood flow between arteries and venous lacunae. Thus, the subepithelially situated capillary network can be avoided, and rapid filling of capacitance veins is possible when the shunts of the arteriovenous anastomoses are open (Paulsen et al. 2000b).

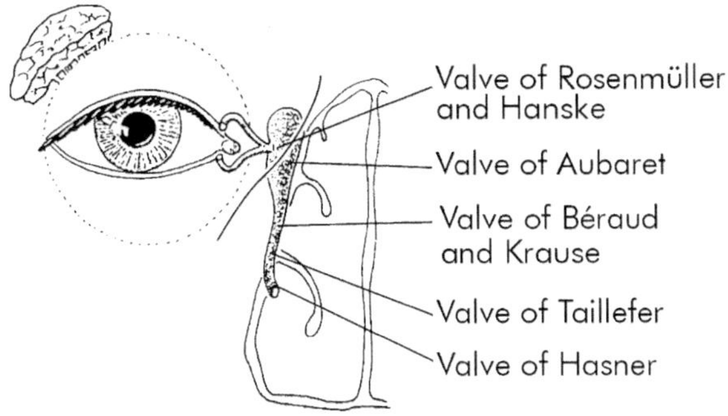

Fig. 28. Schematic view on the valves in the lacrimal sac and nasolacrimal duct described in the past by Rosenmüller, Hanske, Aubaret, Béraud, Krause, Taillefer and Hasner

The occurrence of cavernous tissue in various hollow viscera such as at the entrance to the oesophagus, the uterine tube, the vagina, and the anus is well known (Henle 1866; Stieve 1928). Based on its yielding characteristic, the vascular plexus allows both obstruction and, simultaneously, rapid passage of solid and liquid components.

In the efferent tear duct barrier arteries, capacitance veins and throttle veins facilitate closure and opening of the cavernous body with consecutive regulation of tear outflow.

The possibility has been discussed that tear fluid is absorbed by the epithelial lining before it reaches the nose (Rohen 1964; Paulsen 1998). In this context, the cavernous body could play a role in drainage of the reabsorbed fluid. Moreover, contact times between tear fluid and mucosa may be regulated by the swelling of the cavernous body.

Moreover, it can be assumed that the valves in the lacrimal sac and nasolacrimal duct described in the past by Rosenmüller, Hanske, Aubaret, Béraud, Krause, and Taillefer (Fig. 28) may be based on different swelling states of the cavernous body and must therefore be considered speculative.

Drainage of tears certainly involves a number of different mechanisms. A decisive role is played by capillary attraction (Lesshaft 1868, Frieberg 1917; Rosengren 1928; Jones 1961; Hill et al. 1974; Chavis et al. 1978; Doane 1981), aided by contraction of the lacrimal of the orbicularis muscle with blinking (Rohen 1953; Jones 1958, 1961; Nagashima and Araki 1963; Becker 1992; Thale et al. 1998) and distension of the sac, as well as a passive wringing out of the sac because of its medial attachment and helically arranged fibrillar structures (Thale et al. 1998). The results of the presented study suggest that the cavernous body of the lacrimal passage is the morphologic correlate of a further mechanism that effects tear outflow. When the net outflow of blood from the cavernous body is less than the inflow, the mucosa expands and functionally decreases the tear outflow through the efferent tear duct system. This mechanism acts, for example, to provide protection against foreign bodies that have entered the conjunctival sac; not only is tear fluid production increase by the lacrimal gland, tear outflow is also interrupted by the swelling of the cavernous body to flush out the foreign body and protect the efferent tear ducts themselves (Paulsen et al. 2000d). Moreover, the pathophysiology of functional lacrimal drainage insufficiency (i.e. patients with epiphora despite patent lacrimal passage found during syringing), can be explained by this mechanism: malfunctions in the different blood vessels of the vascular bed may lead to disturbances in the tear outflow cycle, ocular congestion, or total occlusion of the lacrimal passages. Such malfunctions may be caused by acute diseases, such as allergic conjunctivitis, hay fever, or rhinitis, or chronic conditions such as stenosis after dacryocystitis (Paulsen et al. 2001b) or dacryolithiasis. Further, in most patients persistent epiphora after dacryocystorhinostomy can be explained by destruction of the surrounding cavernous body (Paulsen et al. 2000d).

4.2.2
Innervation of the Cavernous Body and the Seromucous Glands

The distribution and function of nervous tissues and neuropeptides in the lacrimal gland (Tsukahara and Tanishima 1974; Seifert et al. 1996), in the accessory lacrimal glands (Seifert et al. 1993, 1997; Seifert and Spitznas 1994), in the meibomian glands (Seifert and Spitznas 1996), and in the nasal vascular bed (Cauna 1970a,b; Cauna and Cauna 1975; Knipping et al. 1995; Riederer et al. 1997) have been investigated extensively.

In human efferent tear ducts, electron microscopy shows the structure and subtle distribution of nerve fibres around subepithelially located seromucous glands and blood vessels of the cavernous body as well as the delicate neuronal profiles in nervous stromal septa between the alveolar buds of seromucous glands and in the connective tissue between blood vessels of the cavernous body. The observation of few axons within the glandular stroma of seromucous glands and around blood vessels of the cavernous body, and especially the presence of nerve fibres with unmyelinated axons, indicate that the seromucous glands and the musculature or the blood vessels of the cavernous body are in fact among the target tissues of these nerve fibres (Paulsen et al. 2000a).

The ultrastructural findings are completed by demonstration of the distributional pattern of different neuronal markers and neuropeptides. RT-97 has been shown to mark an epitope located on the human 200-kDa neurofilament which is found in myelinated axons not only of the central nervous system but also the peripheral nervous system (Wood and Anderton 1981; Anderton et al. 1982; Kahn et al. 1987). S-100 protein, consisting of two subunits, S-100α and S-100β (Haimoto et al. 1987), is a calcium-binding protein first described by Moore (1975) in Schwann cells and other glial cells. In addition to its occurrence in the nervous system, S-100α is also expressed in the exocrine cells of salivary, mammary and sweat glands. Its function in non-nervous tissue is still unknown. It is believed to be involved in the metabolism of free fatty acids and lipolysis in adipocytes (Haimoto et al. 1987). These assumptions are supported by the present finding of S-100 protein in the mucosa of the lacrimal sac and nasolacrimal duct as well as earlier findings of lipid droplets localised in the apical part of the duct epithelial cells (see: Fig. 2e, g; Paulsen et al. 1998). Based on its wide distribution it is supposed that the peptide belongs to the diffuse endocrine cells, the paraneurons (Fujita 1977), which may play a role in the interaction between the endocrine, nervous and immune system in the main lacrimal gland (Kelleher et al. 1991). Using an antibody recognising both subunits, S-100α and S-100β, the protein is identified in the same myelinated nerve fibre bundles marked by RT-97 as well as in finer unmyelinated nerve fibres in the mucosa, the subepithelial region, the blood vessels of the cavernous body, and in the area of the seromucous glands in the efferent tear ducts. Compared with the S-100 protein, immunohistochemistry of neuron-specific enolase permits representation of the whole innervation of the efferent tear ducts. NSE is shown as a molecular marker for nerves of the peripheral and central nervous system (Schmechel et al. 1978).

Neuropeptides are believed to act as transmitters or modulators of transmitters (Cripps and Ptchen-Moor 1989; Cripps and Bennet 1992; Fahrenkrug 1993), Lundberg and co-workers (Lundberg et al. 1980; Johannsson und Lundberg 1981) demonstrated VIP to be co-localise with acetylcholine in many species (Lundberg et al. 1980; Dartt 1989; Matsumoto et al. 1992). VIP is reported to have a direct influence on the stimulation of protein secretion in glandular cells (Dartt et al. 1984). In the human efferent tear ducts, VIP only occurs between the secretory tubules and around blood vessels of the intra- and interlobular connective tissue septa of the seromucous glands, a finding known from other loci in humans and other mammals (Buthler et al. 1984; Sibony et al. 1988; Matsumoto et al. 1992; Seifert et al. 1996; Riederer et al. 1997). SP and CGRP are mainly found in sensory fibres (Robinson et al. 1980; Tervo et al. 1982; Seifert et al. 1996). In some secretory tissues, CGRP has been shown to have a vasodilatory effect with increasing vascular permeability (Lundberg et al. 1988). SP promotes secretomotor and vasodilatory effects (Rudich and Butcher 1976) and protein secretion (Singh et al. 1994). Demonstration of both neuropeptides was found to play a role both in the glandular tissue and the blood vessels of the cavernous body in the efferent tear ducts. NPY is an important peptide in a multitude of physiological processes (Lehmann 1990). Abundantly found in sympathetic neurons, it is used as indicator for this tissue (Lundberg et al. 1982). A similar function can be discussed for TH. TH catalyses the reaction from tyrosine via dopamine to noradrenaline (Lindvall et al. 1981) and thus reveals a sympathetic function. Both neuropeptides could be demonstrated in the efferent tear ducts in a similar manner. They are localised in close association with blood vessels of the cavernous body and seromucous glands (Paulsen et al. 2000a).

The demonstrated neuronal structures and neuropeptides identified in the seromucous glands and the cavernous body are indicative of the complexity of regulation and modulation of the human nasolacrimal duct system by a variety of biological signals (Paulsen et la. 2000a).

The function of the autonomic innervation in the seromucous glands of the efferent tear ducts as well as that of the sensory innervation may thus be in the control of the secretion output. The possibility cannot be ruled out that there is a close structural relationship between the sensory and autonomic fibres in secretory regulation.

The demonstrably dense innervation of the specialised blood vessels of the cavernous body may play an important role in tear outflow by affecting the cavernous body (Paulsen et al. 2000a). It seems to function especially in protection of the efferent tear ducts on the basis of a complex neuronal reflex feedback mechanism starting with the dense innervation of the cornea and ending with the innervation of the cavernous body inside the lacrimal tear ducts.

Besides increased tear secretion from the lacrimal gland and accessory lacrimal glands, epiphora related to emotions such as sorrow or happiness can be explained hypothetically by the same mechanism: cholinergic innervation of specialised blood vessels in the cavernous body can lead to obstruction of the lumen of these blood vessels with a reduction of blood outflow. Consequent swelling of the submucosal cavernous tissue can cause closure of the lacrimal passage. By contrast, adrenergic innervation of the blood vessels of the cavernous tissue may affect a mechanism – albeit controversial – involving relaxation of the submucosal swelling and improvement of lacrimal tear passage. Since the regulation of the cavernous body is subject to autonomic control, a pharmacological influence is a possibility here (Paulsen et al. 2000a).

4.2.3 Experimental Influence on the Regulation of Tear Flow by the Cavernous Body

Thus far, unpublished observation with 30 test persons show that the cavernous body of the nasolacrimal ducts indeed has an influence on tear outflow and reveals pharmacological susceptibility. Interestingly, dropping of xylometazoline on the ocular surface leads after 5 min to prolongation of tear outflow in healthy subjects; in contrast, in test persons with a common cold, tear outflow improves clearly under dropping of xylometazoline. Bringing in a foreign body in the conjunctival sac leads to prolongation of tear outflow in men more than in women and in glasses wearers more than in non-glasses wearers. Provocation by bringing in a foreign body and additional application of xylometazoline also improves tear outflow.

4.2.4 Biomechanics of Tear Flow

The lacrimal sac lined by a double-layered epithelium is composed of a substantia propria arranged in two strata. Duke-Elder (1961) describes underneath the epithelium a layer of adenoid tissue, sometimes aggregated into follicles. Beneath this follows a fibrous layer containing many elastic fibres, supplied by the cavernous body that virtually transforms the layer into erectile tissue continuous with that underlying the nasal mucosa. Besides elastic fibres, the fibrous layer or the lamina propria consists of type I and type III collagen and contains chondroitin-4- and chondroitin-6-sulphate (Paulsen et al. 1998; Thale et al. 1998).

The nasolacrimal duct is composed of a substantia propria and two epithelial layers that closely resemble those of the lacrimal sac (Schaeffer 1922), except for the absence of serous glands in its walls (Iwata 1927). Elastic fibres and collagen bundles between the blood vessels of the cavernous body form a system of connective tissue fibres that is arranged in a helical pattern. In this helical system, fibres run screw shaped from the fornix of the lacrimal sac to the outlet of the nasolacrimal duct (Thale et al. 1998).

Tear flow from the saccus lacrimalis to the cavum nasi is caused by forces that are not completely understood. Some hypotheses propose an active pump mechanism for the function of the human lacrimal sac (Jones 1958; Nagashima and Araki 1963). Fernandez-Valencia and Gomez Pellico (1990) concluded that the saccus lacrimalis is subject to two muscular vectors that are activated during the contraction of the musculus orbicularis oculi. While the muscle contracts, the saccus lacrimalis is dilated, raising the fornix and pulling at the anterior face. When contraction ends, the saccus recovers its original volume.

Other authors suggest that physical factors such as gravity, respiration, absorption and evaporation might play a role in drainage tears through the lacrimal passage (Schirmer 1903; Hurwitz et al. 1975a,b; Murube del Castillo 1978; Nik et al. 1984). Brienen and Snell (1969) postulated that the main, and presumably the sole, force that impels lacrimal flow from the conjunctival sac is the pressure brought about by closing of the eyes; in all probability, their expansions and contractions are secondary consequences of pressure fluctuations in the conjunctival sac. Jones (1958) analysed the anatomical structures of the medial palpebral ligament and the palpebral part of the orbicularis oculi muscle. He introduced the concept of the lacrimal pump system which functions with blinking and might be responsible for lacrimal drainage. Rosengren (1972) showed that during blinking movements the canaliculi are compressed and a uniform volume of lacrimal fluid is squirted into the lacrimal sac. The small canaliculi may act as capillary tubes (Jones 1958; Hurwitz 1996). Lacrimal fluid is attracted by capillarity into the lacrimal puncta, and with closing of the eyelids the contraction of the pre-septal muscle creates a negative pressure and sucks the tear fluid into the sac. The importance of Horner's muscle becomes clear in case of a facial palsy: tears will not be pumped through the lacrimal system. Even with a Jones tube in place, there will be a decrease in tear flow if the orbicularis muscle function is insufficient (Nik et al. 1984). Support for the existence of a canalicular pump system on lid closure also came from experimental work carried out by other workers (Hill et al. 1974). Even the role of gravity is not clear, since Hurwitz (1996) concluded that gravity does have a significant role in the transport of tears. The meaning of the cavernous body for tear outflow has been already discussed above (Sects. 4.1.1–4.1.3; Paulsen et al. 2000a,d).

The structure and organisation of connective tissue in the lamina propria allows a biomechanical consideration of tear flow within the meaning of functional adaptation: With blinking, the lacrimal part of the orbicularis muscle contracts. The fornix of the sac moves in a cranial-lateral direction. Thus the lacrimal sac distends and may be "wrung out" passively because of its medical attachment and helically arranged fibrillar structures (Fig. 19; Thale et al. 1998). The capillary bed of the cavernous body that is embedded in the helical system may play an important role in absorption and drainage of lacrimal fluid (Thale et al. 1998).

4.3 Absorption of Tear Fluid Components

4.3.1 What Happens to Tears Inside the Nasolacrimal Ducts?

From each drop applied to any eye, only a small portion is taken up into the eye by the conjunctiva, sclera or cornea. A larger part is drained via the upper and lower lacrimal puncta into the nasopharyngeal space, where an enlarged mucosal surface is available for absorption. The remaining part is swallowed, absorbed in the gastrointestinal tract, and may be metabolised during liver passage, whereas drugs absorbed in the nasopharynx are not subjected to this mechanism. They bypass the liver to enter systemic circulation directly. The administration of eye drops has therefore been compared to a slow intravenous injection. Compression of the lacrimal puncta can reduce systemic absorption considerably (Passo et al. 1984). This concept disregards an important issue: What happens to the applied teardrop inside the nasolacrimal ducts, and what happens to tears inside the efferent lacrimal passage?

Tucker and Codere (1994) showed the median dye transit time of an applied tear drop containing fluorescein to be 8 min. until its appearance in the nose. As early as 1964, Rohen suggested absorption of tear fluid through the lining epithelium of the human efferent lacrimal passage. Current studies of the human nasolacrimal ducts could confirm this speculation. It was demonstrated that the epithelial cells of the human lacrimal sac and nasolacrimal duct are double-layered in most areas and characterised by microciliation (Paulsen et al. 1998; Thale et al. 2001). Moreover, human nasolacrimal ducts are surrounded by a cavernous body (Paulsen et al. 2000a,d). It may enable closure and opening of the lumen of the lacrimal passage by swelling and shrinkage of the cavernous body with consequent regulation of tear outflow (Paulsen et al. 2000d). The retention time of tear fluid in the efferent tear ducts may be prolonged by this mechanism. Also, formation of dacryoliths in the human efferent tear ducts may be a hint of the possible role of absorption in the lacrimal sac and nasolacrimal duct (Maltzman and Favetta 1979; Yazici et al. 2001). Therefore, we tested the hypothesis of Rohen (1964) – as to whether absorption of tear fluid in the efferent tear ducts takes place or not – in an animal experiment.

Absorption of macromolecular substances and their uptake and transportation into the blood and lymph are controlled by morphological and functional barriers. Functional barriers include enzymes of the epithelium that could destroy macromolecular substances before they can be absorbed, and the intrinsic properties of the macromolecules including solubility, lipophilicity and charge. The morphological barriers are represented by the anatomical structures of the mucosa (Gebert 1995).

The purpose of this research was to study the hypothesis of Rohen (1964), concerning whether parts of the tear fluid or tear fluid components are absorbed by the lining of the efferent lacrimal passage before reaching the nose. In rat we could show that proteins cannot cross the morphological and functional barriers of the efferent tear ducts as in intact molecules. This may be different in humans because of an unexpected different morphology. Although the occurrence of intraepithelial mucus glands is very similar between both species, the multilayered epithelium of the rat efferent tear ducts with larger squamous cells above and cuboidal cells below makes absorptive processes of macromolecules nearly impossible. This finding correlates to the finding that only traces of applied iodinated BSA were incorporated into the circulation of the rat. It cannot be excluded with certainty that the measured 0.13% of the applied radioactivity was a product of absorption across the respiratory epithelium of the rat nose. An even higher part of the radioactivity was taken up as the proteolytic degradation product of BSA to free amino acids, because this molecule is cleaved fairly rapidly by mucosal enzymes (high functional barrier). Ninety-six percent of the radioactivity incorporated after 60 min. was free iodine, probably as a contaminant of iodinated preparation.

The failure of a large globular protein like albumin to be taken up does not mean that peptides or other organic compounds may not be taken up. Regarding the free label, Maurice (1955, 1967, 1973) as well as Macdonald and Maurice (1991) showed that a label is absorbed across the conjunctival surface and the corneal surface. Thus, having a high level of free label in the present study it is surprising that this is not taken up by the epithelium of the lacrimal sac and duct.

It can be concluded that the rat is unfit for studying absorptive processes in the efferent tear ducts.

4.3.2 The Nasolacrimal System of the Rabbit Shows High Homology with the Human Nasolacrimal Ducts

Based on the finding that the rat is unsuitable for studying absorptive processes in the nasolacrimal ducts (see above; Paulsen et al. 2000c), and as there were practically no further data available on vertebrate histology of the efferent tear ducts, we first performed comparative anatomy procedures to determine which animal species could serve for further absorption experiments.

The present comparative findings demonstrate that the nasolacrimal system of the rabbit is most suitable for absorptive studies as it reveals the highest homology with the ape nasolacrimal system of all animals compared and also a very high homology with the human nasolacrimal system, except the distribution and integration of goblet cells in the epithelium (Table 5). Moore et al. (1987) showed that the dog is most suitable for studies about the mucous system, as there are high similarities between canine and human conjunctival goblet cell distribution. Unfortunately, dogs were not available for our comparative studies. Nevertheless, the rabbit allows inferences concerning absorptive processes that may take place in human efferent tear ducts.

4.3.3 The Rabbit – Animal Model for the Absorption of Lipophilic Substances from Tear Fluid by the Epithelium of the Nasolacrimal Ducts

The purpose of our research was to ascertain whether, as suggested by Rohen (1964), the lining of the efferent lacrimal passage is able to absorb parts of the tear fluid or tear fluid components before reaching the nose. To answer this question in an animal experiment, comparative anatomy was conducted, since the absorption experiment in rats with iodinated albumin showed that the rat is unsuitable for absorptive processes in the efferent tear ducts (see above; Paulsen et al. 2000c).

In the absorption experiment with rabbits, we demonstrated that lipophilic steroids can cross the morphological and functional barriers of the nasolacrimal ducts in rabbits. Although the morphology of the lining epithelium and the surrounding cavernous body of the nasolacrimal ducts are very similar in rabbits and primates as shown in the comparative study, absorptive processes in the nasolacrimal ducts of the rabbit are not transferable to humans without further ado. It is well known that the rabbit has a very low eye blink rate with an average of one blink every 20–30 min. (Gormezano et al. 1962) and they can also manage for long periods without drinking water (Tarjan et al. 1984; Denton et al. 1985), suggesting special enabling mechanisms. Maurice (1995) questioned the effect of the low blink rate in rabbits on topical drug penetration through the cornea and showed that for most drugs the epithelial permeability of the cornea is so high that all the material present in the tear film will partition into the cornea in a few minutes. Thus, he concluded, considerations of contact time become irrelevant. Also for humans, both Maurice (1955, 1967, 1973) and Macdonald and Maurice (1991) showed that a labelled substance is absorbed across the conjunctival surface and the corneal surface. However, the rate of loss in humans seems to be much lower than in rabbits, leading to an average overestimate of tear turnover of 25% (Macdonald and Maurice 1991). Although we largely avoided the effect of ^{3}H-cortisol loss through the cornea and conjunctiva by closing the rabbit eyes manually several times after dropping the substance and also showed by autoradiography of the nasolacrimal ducts that ^{3}H-cortisol is absorbed through the lining of the efferent tear ducts in the upper part of the nasolacrimal duct as well as in the lower part, partial absorption of the measured radioactivity by the nasal mucosa cannot be excluded. However, as demonstrated in the present study by SEM and supported by the previous finding that the epithelium of the human nasolacrimal sac and nasolacrimal duct possess microciliation (Paulsen et al. 1998), our experiment shows that absorption takes place in the nasolacrimal passage. Contrary to the case in rabbits, the eye blink rate in adult humans is many times higher (averaging 17/min.). Thus, most of the tear fluid as well as applied substances such as drugs enter the nasolacrimal ducts, where they have an average 8 min. to be absorbed as shown by Tucker and Codere (1994).

Our data suggest that the rabbit can serve as an animal model for the analysis of absorption of substances form tear fluid by the epithelium of the nasolacrimal ducts (Paulsen et al. 2002e). Such investigations can include normal tear components, such as freely water-soluble small molecules (e.g. urea or amino acids) or perhaps smaller model tear proteins such as lysozyme. These investigations would be useful for extrapolation on the human situation as an understanding of the exact mechanism of absorption and regulation of these processes at the mucosa of the lacrimal passage could provide further insights into the occurrence of dacryoliths, which often lead to residual functional impairment with epiphora. Moreover, some factors such as old age, changes in hormonal status (postmenopausal women), immunodeficiency or the sicca syndrome could be associated with downregulation of transporter systems necessary for absorption, and the hypothesis could be analysed concerning whether the normally constant absorption of tear fluid components into the blood vessels of the surrounding cavernous body (Paulsen et al. 2000c, 2000d) that are connected to the blood vessels of the ocular surface (Paulsen et al. 2000c) could be a feedback signal for tear fluid production.

It can be concluded that the epithelium of the nasolacrimal passage of the rabbit is able to absorb lipophilic substances, and it is most likely that this is also true for the epithelium of the human lacrimal sac and nasolacrimal duct. We therefore suggest that lipophilic components of tear fluid or drugs from tear drops are also readily absorbed by the nasolacrimal ducts, by-passing primary metabolism in the liver, and thus they can display direct systemic effects. The absorption of cortisol in the rabbit nasolacrimal ducts and the similarities which exist between rabbit and human nasolacrimal ducts support the use of the rabbit as a model species for studying absorption of substances from tear fluid in the nasolacrimal ducts.

4.4 Insights into the Pathophysiology of Primary Acquired Dacryostenosis

PANDO results from inflammation of unknown cause that eventually leads to occlusive fibrosis (Bartley 1992). It was observed that the overall incidence of lacrimal obstruction on one or both sides was more or less equal in the two sexes and directly proportional to age, and that 35–40 years was the earliest expected age at onset of an obstruction of this type (Dalgleish 1967). Today, it is generally accepted that idiopathic dacryostenosis is seen more often in females, especially in post-menopausal women (Schnyder 1920; Meller 1929; Traquair 1941; Summerskill 1956; Ruiz Barranco 1966; Busse and Müller 1977; Zolli and Shannon 1982; Linberg and McCormick 1986; Tarbet and Custer 1995). Our present data underline these findings with a female/male ratio of 30/6 and a median age of 55.5 years, at which there were three females aged below 30.

The role of the osseous nasolacrimal canal in primary acquired dacryostenosis has been discussed frequently, usually arriving at the suggestion that a narrow canal predisposes a person to development of lacrimal obstruction (Meller 1929; Jones and Boyden 1972; Steinkogler 1986; Groessl et al. 1997; Janssen 2001). Also, the canal is said to be longer and narrower in women than in men, and longer in Caucasians than in other human sub-groups (Santos-Fernandez 1921; Meller 1929). Furthermore, a familial disposition and osteoporotic changes have been proposed as factors (Duke-Elder 1974; Leone 1986). On the other hand, remarkable variations have been reported in the canal in a series of normal skulls (Whitnall 1912; Rheeman 1998), suggesting that stenosis of the osseous canal is not the cause of nasolacrimal duct obstruction.

Congenital diverticula were observed by Melanová (1969), who considered them to be the cause of primary acquired dacryostenosis, whereby collections of secretions and foreign bodies in the blind sac were said to lead to chronic inflammation. However, diverticula of the lacrimal passage are rare and their opening is normally directed against the nose so as to oppose the collection of secretions (Busse and Müller 1977).

Although the antimicrobial effect of the tear fluid is well known from anatomical and immunologic studies, the whole efferent pathway seems to function mainly as a contributor to antimicrobial defence, in addition to tear drainage (Milder 1983; Perra et al. 1995; Paulsen 2001; Paulsen et al. 1998, 2000b; 2001a–c; 2002b–d; Sirigu et al. 2000; Knop and Knop 2001). Descending infection from the conjunctiva has been considered as a causative factor in dacryostenosis (Summerskill 1956). The obstruction is believed to occur as a response to tiny foreign bodies and pathogens, leading to infection and swelling of the nasolacrimal duct. Other studies have implicated sinus and nasal disease as a causative factor in nasolacrimal duct obstruction. Because the canal is within the medial wall of the maxillary sinus and opens into the nasal cavity, it would seem vulnerable to secondary involvement in sinus or nasal disease. Early reports associated sinusitis or nasal disease with 20%–55% of nasolacrimal obstructions (Traquair 1941; Garfin 1942; Dorello 1957). However, clinical studies indicate that nasal disease is extremely rare in patients undergoing DCR (Lingerg and McCormick 1986). Also, the cohort presented here, with two patients revealing signs of sinusitis and two revealing nasal disease, suggests that sinus and lacrimal disease are not associated.

Interestingly, the mucous membranes of the lacrimal passage and nose reveal morphologic differences between the nasolacrimal epithelial cells with microciliation (Paulsen et al. 1998) only and the nasal epithelial cells with their kinocilia. This also suggests differences in susceptibility to pathogens. However, it has been shown – and our TEM results underline – that ectopia of nasal epithelial cells is a more or less common finding in the nasolacrimal duct (Radnot 1977; Adenis et al. 1980). In addition to descending infection from the eye, ascending infection of these atypical cells during nasal inflammation could thus be the starting-point of dacryostenosis. In this scenario, the nasal inflammation has long since abated when the dacryocystitis passes into a chronic state, causing the changes described.

Our present findings are in accordance with previous histopathologic descriptions of primary acquired dacryostenosis (Busse and Müller 1977; Linberg and McCormick 1986). All these papers show epithelial changes as well as fibrosis of the lamina propria and conclude that these changes are causative for the obstruction. However, previous authors were unable to explain cases in which the nasolacrimal duct remained mechanically patent to forced irrigation. These patients with so-called functional obstruction had constant epiphora, but reported no history of acute or chronic dacryocystitis (Linberg and McCormick). Digital pressure over the sac and irrigation failed to produce mucopurulent reflux. According to Rosenstock and Hurwitz (1982) the term "functional obstruction" is confusing and should be avoided. They show by intubation dacryocystography with subtraction that patients with tearing whose lacrimal drainage passages are patent to syringing may have an incomplete anatomic obstruction. Alternatively, patients with tearing whose lacrimal drainage passages are patent to syringing may have anatomically normal lacrimal passages but a physiologic dysfunction that can be demonstrated by nuclear scintillography, or with computer assistance when the abnormality is subtle. In such cases, the authors believe the problem usually involves the eyelids, punctum or lacrimal pump, though in rare cases a lacrimal sac that drains poorly is at fault.

At any rate, our cohort clearly demonstrates that "functional obstruction" of the lacrimal passage occurs in many more patients than suggested by Linberg and McCormick (1986). Furthermore, some of these patients report a history of acute or chronic dacryocystitis (Table 6: cases 1, 6, 13, 18, 21, 22, 24) and digital pressure over the sac also revealed mucopurulent reflux in cases 13 and 22.

Linberg and McCormick (1986) presume that functional obstruction initially occurs secondary to soft tissue compression of the duct by way of a chronic inflammatory process that antedates the onset of infection. Busse and Müller (1977) hypothesise that the surrounding vascular plexus of the nasolacrimal duct could play a role in dacryostenosis. This vascular plexus was first described by Henle (1866) in 1866 and a recent publication showed that this vascular plexus is comparable to a cavernous body (Paulsen et al. 2000a,c,d). While regulating the blood flow, the specialised blood vessels permit opening and closing of the lumen of the lacrimal passage, effected by the bulging and subsiding of the cavernous body, and at the same time regulate tear outflow. In this context, the description of nasolacrimal duct obstruction caused by topical epinephrine abuse (Spaeth 1967) could indicate drug susceptibility of the surrounding cavernous body. The cavernous body is embedded in a system of collagen bundles, elastic and reticular fibres arranged in a helical pattern (Thale et al. 1998). In this helical system, fibres run screw-shaped from the fornix to the outlet of the nasolacrimal duct. Besides the cavernous body, the arrangement of subepithelial connective tissue is said to play a decisive role in the tear outflow mechanism; based on the contraction of the lacrimal part of the orbicularis muscle during blinking, the fornix of the sac moves in a cranial-lateral direction. Thus the lacrimal sac distends and the lacrimal passage may be "wrung out" passively because of its medial attachment and helically arranged fibrillar structures (Thale et al. 1998).

In addition to squamous metaplasia and subepithelial fibrosis with remodelling of the helically arranged connective tissue fibres, our results demonstrate that primary acquired dacryostenosis results in reduction and even total loss of specialised blood vessels of the surrounding cavernous body and their replacement by connective tissue, suggesting an interruption of the tear flow mechanism in the stenotic segment. Taking into consideration the literature and our present findings, we postulate the following pathogenic concept of primary acquired dacryostenosis: Descending inflammation from the eye or ascending inflammation from the region of the nose (perhaps by way of inflammation of "ectopical" nasal epithelial cells in the nasolacrimal duct) may initiate malfunctions in the cavernous body with reactive hyperaemia, swelling of the mucous membrane and temporary occlusion of the lacrimal passage. Then, repeated isolated episodes of dacryocystitis may lead to the structural epithelial and subepithelial changes presented. Loss of typical goblet and epithelial cells, which play an important role in tear outflow mechanism (Perra et al. 1995), as well as fibrosis of the helical system of connective tissue fibres in the area of the lacrimal sac and nasolacrimal duct and reduction and destruction of specialised blood vessels of the cavernous body, may exacerbate malfunctions of the tear outflow mechanism and start up a vicious circle (Duke-Elder 1952).

Thus, cases of functional dacryostenosis, i.e. patients with epiphora in spite of patent lacrimal passages on syringing, are explainable as follows: (1) In early cases, the rinsing liquid instilled by hand pressure into the lacrimal passage is forced through the segment of the lacrimal passage obturated by reactive hyperaemia and swelling of the mucous membrane; (2) advanced structural changes (Table 6: cases 13, 21, 22, 24) caused by chronic dacryocystitis result in a non-functioning segment in the lacrimal passage. In this latter case, obstruction is not yet complete, but there is no tear transport in the non-functioning segment.

Ample evidence certainly exists that obstructed lacrimal systems are colonised by increased numbers of pathogenic micro-organisms (Traquair 1941; Sood et al. 1967; Sasaki et al. 1973; Blicker and Buffan 1993). The histopathology implicates that, besides topical or systemic anti-inflammatory agents, nasal decongestants such as Afrin possibly could be also useful tools in the management of early cases of primary acquired dacryostenosis, since they could counteract the hyperaemia and swelling of the cavernous body. Further investigations are planned to evaluate the function of the cavernous body and the effect of nasal decongestants on this tissue.

A detailed understanding of the pertinent anatomy and pathophysiology of the nasal portion of the lacrimal outflow system gives the ophthalmologist an overview of both sides of the problem, and perhaps equips him best to orchestrate both ophthalmic and nasal management of lacrimal outflow disorders.

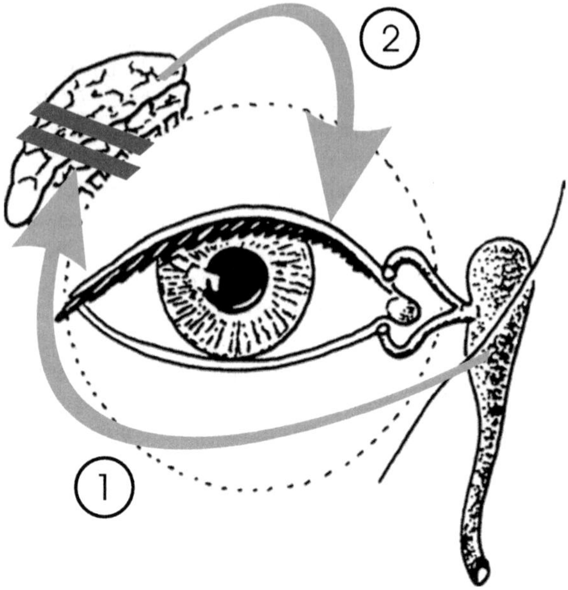

Fig. 29. The normally constant absorption of tear fluid components into the blood vessels of the surrounding cavernous body of the nasolacrimal ducts and their transport to the lacrimal gland by blood vessel connections (*1*) could be a feedback signal for tear fluid production (*2*)

4.5 Considerations on the Relationship Between Dry Eye and the Nasolacrimal Ducts

4.5.1 Absorption of Tear Fluid in the Nasolacrimal Ducts – Feedback Signal for Tear Fluid Production?

There has been a considerable increase in knowledge regarding the pathogenesis of the keratoconjunctivitis sicca that occurs in dry eye disease. The pathologic features of this condition include increased epithelial stratification and proliferative index and abnormal differentiation with maintenance of a basal phenotype (Jones et al. 1998). Furthermore, there is reduced expression of secretory and membrane-bound mucins by the superficial ocular surface epithelial cells (Pflugfelder et al. 1997; Danjo et al. 1998).

An exact mechanism for the development of these pathologic changes has not been elucidated. The severity of keratoconjunctivitis sicca worsens as aqueous tear secretion decreases and as the ability to reflex tear in response to sensory stimulation is lost. The disease lead to a vicious cycle (Pflugfelder et al. 1998, 2000).

As already discussed and described above, our data suggest that the human nasolacrimal ducts have absorptive functions. Based on this, it could be suggested that the normally constant absorption of tear fluid components into the blood vessels of the surrounding cavernous body (Paulsen et al. 2000a,d) that are connected to the blood vessels of the ocular surface (Paulsen et al. 2000d) could be a feedback signal for tear fluid production, which comes to a halt when these tear components are not absorbed (Fig. 29).

As this hypothesis is true, one must ask naturally, why not all possible forms of nasolacrimal diseases that are related with tear outflow disturbances lead to dry eye? And why do punctum plugs and other methods that interrupt the connection between the ocular surface and the nasolacrimal ducts sometimes show very good results in the therapy of dry eye? The following can be answered: (1) Currently, nothing is known of how diseases of the drainage system coincide with changes in tear fluid production, although patients suffer from epiphora. (2) Both dry eye and diseases of the nasolacrimal ducts occur with high frequency in women of middle-to-advanced age (McCarty et al. 1998). (3) Looking the results after insertion of punctum plugs and other blocking methods, one can ascertain that patients – viewed over a longer period – practically do not manage without additional tear substitutions. Punctum plugs could in the first instance additionally function as they totally prevent absorption of tear fluid components. An "empty" tear fluid system may be a strong stimulation signal for tear fluid production. This stimulation signal is decreased time-dependent. Also, in cases of dacryostenosis or total closure of the nasolacrimal passage in the area of the canaliculi, the epiphora may be still strengthened.

It is recognised today that the ocular surface (cornea, conjunctiva, lacrimal gland, accessory lacrimal glands, and Meibomian glands) and the inter-connecting neural-reflex arcs function as an integrated or "functional unit (Stern et al. 1998). The human nasolacrimal ducts are integrated in this reflex arc as shown by our investigations (Paulsen et al. 2000a; F. Paulsen, unpublished results). Disturbances in one part of this complex regulatory system can have effects for all participant structures.

4.5.2
Tear Duct-Associated Lymphoid Tissue – Entrance Side for the Immunological Events in Dry Eye?

Some organs of the human body (anterior eye chamber, brain, placenta, testicle) reveal a special immunological state of reduced activation of the specific and unspecific immune system. This condition of local immune suppression that is termed *the immune privilege* expresses in delayed or totally suppressed rejection of allogenic transplantations in these organs (Medawar 1948) as it is illustrated in the remaining of the immunophenotypic childish placenta in the mature organism, but also in the survival of corneal and lens transplants into the anterior eye chamber. The biological functions of the immune privilege are evident: tolerance of a foreign antigen obviously is better in some organs than its elimination, which only can be reached for the price of T cell-mediated cytolysis of local cells. Such cell loss is not compensable in badly regenerative, postmitotic or highly differentiated tissues. Therefore, some viruses survive in the central nervous system, as their elimination by T-effector cells unquestionably would lead to neural cell death with severe neurological deficit or individual death. This is similarly valid for the anterior eye chamber (Streilein 1996) and the testicle. Such an immune suppression is not necessary in regenerative organs like the liver or the skin, as all cells needed for this process are able to proliferate and redifferentiate.

The mechanisms that maintain the immune privilege are non-uniform for the different organs and are not enlightened in detail. Beside mechanical tissue barriers as a classical conception (i.e. bood–brain, blood–testis, or blood–retina barrier)(Medawar 1948), the expression of so-called death-ligands (CD95, TRAIL, TNF), that induce apoptosis of potentially dangerous T cells (Bechmann et al. 1999a,b, 2000; Bellgrau et al. 1995; Griffith et al. 1995; Nitsch et al. 2000) are also under discussion, as well as a special form of antigen presentation which produces immune tolerance (Streilein and Niederkorn 1985; Bechmann and Nitsch 2000; Bechmann et al. 2000; Brabb et al. 2000; Wenkel et al. 2000). Such *immune deviation* was described first in the anterior eye chamber (Streilein and Niederkorn 1985). Here, injection of foreign antigen does not lead – as at other locations of the body – to a local T-cell reaction (type IV immune reaction) but produces systemic tolerance against the inoculated antigen. In this way, antigens are not attacked in the anterior eye chamber, and thus the sensitive visual system will be protected against inflammatory damage. As described above, the immune privilege of the anterior eye chamber also allows transplantation of allogenic lenses, artificial intraocular lenses, and corneas (nevertheless type IV immune reactions are possible after cornea transplantation in rare cases).

It is known that the tolerance is adoptive-transferable by injection of splenocytes from one animal that was primed by inoculation with an antigen into a second animal. This result (Streilein and Niederkorn 1985) demonstrates that antigens from the anterior eye chamber are presented with a signal that produces immune deviation, and that regulatory T cells have developed. In contrast to the spleen, the cervical lymph nodes do not play a critical role in the induction of immune deviation, as was shown in rats in an actual study by Yamagami and Dana (2001). Nevertheless, the drainage ways of the antigens from the anterior eye chamber, the location of their presentation, as well as the passage of the belonging antigen-presenting cells are unknown at present. It is especially unclear which role is played by the conjunctiva and the nasolacrimal ducts as well as their associated lymphoid tissues [CALT (Axelrod and Chandler 1979, Chodosh et al. 1998; Dua et al. 1994; Knop and Knop 2000; Osterlind 1944; Wotherspoon et al. 1994) and TALT (Paulsen et al. 2000b; Paulsen et al. 2002b; Knop and Knop 2001)] for the immune privilege of the anterior eye chamber.

As already described, our group has shown the presence of TALT in human (Paulsen et al. 1998, 2000b, 2002b). Based on this late description of TALT, it has not been taken into consideration until now that antigens drained by the tear fluid could by themselves induce immune deviation. With regard to protection of the cornea against inflammatory destruction, this would be plausibly analogous to the nervous system (Bechmann et al. 2001; Brabb et al. 2000; Wenkel et al. 2000) and the anterior eye chamber (Streitlein and Niederkorn 1985).

Activation of T lymphocytes has been observed in dry eye which leads to the frequent occurrence of abnormal (pathological) apoptosis in terminally differentiated, acinar epithelial cells of the lacrimal gland (Gao et al. 1998). Tears, now secreted to the ocular surface, will contain proinflammatory cytokines and will inflame the tissues of the ocular surface. Abnormal apoptosis has also been detected within the epithelial cells and lymphocytes of the ocular surface (Gao et al. 1998). This ocular surface inflammatory response consists of inflammatory cell infiltration, activation of the ocular surface epithelium with increased expression of adhesion molecules, inflammatory cytokines and pro-apoptotic factors, increased concentrations of inflammatory cytokines in the tear fluid, and increased activity of matrix-degrading enzymes in the tear fluid (Pflugfelder et al. 2000). Hints exist that the reduction of circulating androgens plays a role for these processes (Azzarolo et al. 1999; Sullivan et

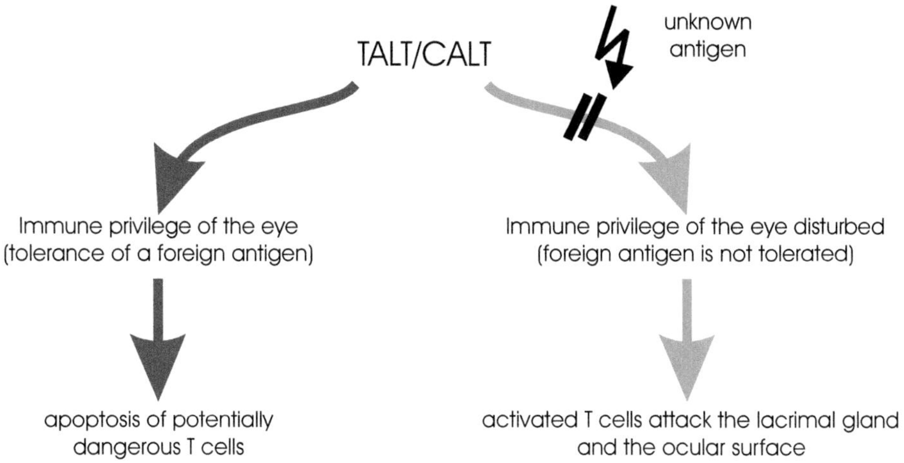

Fig. 30. Possible meaning of TALT/CALT in the pathogenesis of dry eye

al. 1999). Recently, it was shown that treatment with locally applied cyclosporin A eye drops intervents the interleukin (IL) metabolism, especially of IL6, and thus creates a treatment option for the first time which leads to a remarkably improvement of the irritation symptoms and ocular surface signs, especially in severe cases of keratoconjunctivitis sicca.

All these findings lead to the adoption that TALT and also CALT may play a role in the pathogenesis of dry eye. One can imagine that misdirected stimulation of TALT and/or CALT can result in a misled form of immune deviation at the ocular surface. Within the scope of this event, T cells are not hindered any longer by apoptosis to induce autoimmunity, which finally results in the frame of dry eye (Fig. 30).

5 Outlook

The findings obtained by our group during recent years show that the human nasolacrimal ducts must be included in the considerations concerning dry eye. Interesting areas will be analyses of normal tear components, such as freely water-soluble small molecules (e.g. urea or amino acids) or perhaps smaller model tear proteins such as lysozyme. These investigations would be useful for extrapolation to the human situation, since an understanding of the exact mechanism of absorption and regulation of these processes at the mucosa of the lacrimal passage is still missing. In this context, the current investigation of epithelial transporter systems of the human nasolacrimal ducts will give deeper insights into possible routes of absorption and also the substances which are able to be absorbed. Moreover, current analysis of TALT, CALT, as well as different epithelial cells of the lacrimal passage will be very interesting concerning the immune privilege of the eye.

6 Summary

The human nasolacrimal ducts are a borderland of the disciplines ophthalmology and otorhinolaryngology, which work close together in the treatment of nasolacrimal disorders. However, little knowledge exists concerning the physiology of the nasolacrimal system and, therefore, its pathophysiology. The aim of the present investigations was to obtain a better understanding of tear flow physiology, microbial defence strategies and pathophysiological processes by analysing the nasolacrimal system, thus opening new perspectives against existing therapeutical concepts of diseases of the nasolacrimal ducts.

The human nasolacrimal ducts consist of the upper and the lower lacrimal canaliculus, the lacrimal sac and the nasolacrimal duct. They drain the tear fluid from the ocular surface into the lower meatus of the nose. The lining epithelium of the lacrimal sac and the nasolacrimal duct is faced by microvilli, and animal experiments in rabbits support the hypothesis that tear fluid components are absorbed in the human nasolacrimal system. Evidence of seromucous glands, intraepithelial lipids and a specific mucus layer leads to the conclusion that the normal three-layering of the tear film is also present in the nasolacrimal ducts. Based on its composition, the mucus layer, which is mainly synthesised by goblet cells and intraepithelial mucus glands, functionally serves as a simplified drainage of tear fluid, and it is attributed similar properties to epithelia of the gastrointestinal tract. Further defence mechanisms are represented by antimicrobial peptides IgA and immunocompetent cells (lymphocytes and macrophages) which show intraepithelially and subepithelially a special distribution. Moreover, organised lymphoid tissue is present, which reveals the cytomorphologic and immunophenotypic characteristics of mucosa-associated lymphoid tissue.

The wall of the lacrimal sac and the nasolacrimal duct are made up of a helical system of different connective tissue fibres. Wide luminal vascular plexus are embedded in this helical system, which is comparable to a cavernous body. Caudally, the vascular system is connected to the cavernous body of the inferior turbinate. With distension the system may be "wrung out" due to its medial attachment and helically arranged fibrillar structures. Thereby, tear fluid is drained distally. The embedded blood vessels underlie vegetative control. By means of this innervation, the specialised blood vessels permit regulation of blood flow by opening and closing the lumen of the lacrimal passage as effected by the engorgement and subsidence of the cavernous body, at the same time regulating tear outflow. Related functions, such as a role in the occurrence of epiphora related to emotional responses, are relevant. Moreover, malfunctions in the cavernous body and in its innervation may lead to disturbances in the tear outflow cycle, ocular congestion or total occlusion of the lacrimal passage.

Thus, the pathophysiology of primary acquired dacryostenosis can be explained: Descending inflammation from the eye or ascending inflammation from the nose initiates swelling of the mucous membrane, remodelling of the helical arrangement of connective tissue fibres, malfunctions in the subepithelial cavernous body with reactive hyperaemia, and temporary occlusion of the lacrimal passage. In the follow-up, repeated isolated occurrence of dacryocystitis leads to structural epithelial and subepithelial changes, which may lead either to a total fibrous closure of the lumen of the efferent tear duct or to a non-functional segment in the lacrimal passage that is manifest on syringing.

The normally constant absorption of tear fluid components into the blood vessels of the surrounding cavernous body that are connected to the blood vessels of the outer eye could be a feedback signal for tear fluid production, which comes to a halt if these tear components are not absorbed. Thus, dry eye could be initiated. Defective stimulation of tear duct-associated lymphoid tissue (TALT) could result in abnormal immune deviation at the ocular surface leading to an autoimmunologic response that could cause dry eye pathology.

References

Abraham SN, Beachey EH (1985) Host defenses against adhesion of bacteria to mucosal surfaces. In: Gallin JF, Fauci AS (eds) Advances in host defense mechanisms. Vol 4. Raven press, New York, pp 63-88

Adenis JP, Loubet A, Leboutet MJ, Robin A, Loubet R (1980) Morphologie ultrastructurale de la muqueuse des voies lacrymalis à ses différents niveaux. Arch Anat Cytol Pathol 28: 371-375

Alison MR, Chinery R, Poulsom R, Ashwood P, Longcroft JM, Wright NA (1995) Experimental ulceration leads to sequential expression of spasmolytic polypeptide, intestinal trefoil factor, epidermal growth factor and transforming growth factor alpha mRNAs in rat stomach. J Pathol 175:405-414

Alpan O, Rudomen G, Matzinger P (2001) The role of dendritic cells, B cells, and M cells in gut-oriented immune responses. J Immunol 166:4843-4852

Anderton BH, Breinburg D, Dowens MJ, Green PJ, Tomlinson BE, Ulrich J, Wood JN, Kahn J (1982) Monoclonal antibodies show that neurofibrillary tangles and neurofilaments share antigenic determinants. Nature 298:84-86

Arlt v F (1855) Über den Thränenschlauch. Arch Ophthalmol I(2):135-160

Axelrod AJ, Chandler JW (1979) Morphologic characteristics of conjunctival lymphoid tissue in the rabbit. In: Silverstein AM, O'Connor GR (eds) Immunology and Immunopathology of the Eye. New York: Masson pp 292-301

Azzarolo AM, Wood RL, Mircheff A, Richters A, Olsen E, Berkowitz M, Bachmann M, Huang ZM, Zolfagari R, Warren DW (1999) Androgen influence on lacrimal gland apoptosis, necrosis, and lymphocytic infiltration. Invest Ophthalmol Vis Sci 40:592-602

Babyatsky MW, de Beaumont M, Thim L, Podolsky DK (1996) Oral trefoil peptides protect against ethanol- and indomethacin-induced gastric injury in rats. Gastroenterology 110:489-497

Bains RM, Miller KD (1988) Peroxidase labelling in immunocytochemistry, a critical comparison of 5 systems. Med Lab Sci 45: 240-244

Bals R, Wang X, Wu Z, Freeman T, Bafna V, Zasloff M, Wilson JM (1998) Human beta-defensin 2 is a salt-sensitive peptide antibiotic expressed in human lung. J Clin Invest 102:874-880.

Bartley GB (1992) Acquired lacrimal drainage obstruction: an etiologic classification system, case reports, and a review of the literature. Part 1. Ophthal Plast Reconstr Surg 8:237-42

Bechmann I, Lossau S, Steiner B, Mor G, Gimsa U, Nitsch R (2000) Reactive astrocytes upregulate Fas (CD95) and FasL (CD95L) expression, but do not undergo programmed cell death during the course of anterograde degeneration. Glia 32:25-41

Bechmann I, Mor G, Nilsen J, Eliza M, Nitsch R, Naftolin F (1999) FasL (CD95L, Apo1L) is expressed in the normal rat and human brain: evidence for the existence of an immunological brain barrier. Glia 27:62-74

Bechmann I, Nitsch R (2000) Involvement of non-neuronal cells in entorhinal-hippocampal reorganization following lesions. Ann NY Acad Sci 911:192-206

Bechmann I, Peter S, Beyer M, Gimsa U, Nitsch R (2001) Presence of B-7.2 (CD86) and lack of B7-1 (CD80) on myelin-phagocytosing MHC-II positive rat microglia is associated with non-destructive immune response in vivo. FASEB J 15:1086-1088

Becker BB (1992) Tricompartment model of the lacrimal pump mechanism. Ophthalmology 99:7

Bellgrau D, Gold D, Selawry H, Moore J, Franzusoff A, Duke RC (1995) A role for CD95 ligand in preventing graft rejection. Nature 377:630-632

Berry M, Ellingham RB, Corfield AP (1996) Polydispersity of normal human conjunctival mucins. Invest Ophthalmol Vis Sci 37:2559-2571

Bienenstock J (1974) The physiology of the local immune response in the gastrointestinal tract. Prog Immunol 4:197-209

Blicker JA, Buffan FV (1993) Lacrimal sac, conjunctival, and nasal culture results in dacryocystorhinostomy patients. Ophthalmic Plast Reconstr Surg 9:43-46

Boukamp P, Petrussevska RT, Breitkreutz D, Hornung J, Markhan A, Fusenig NE (1988) Normal keratinization in a spontaneously immortalized aneuploid human keratinocyte cell line. J Cell Biol 106:761-771

Bowman H (1995) Peptide antibiotics and their role in innate immunity. Ann Rev Immunol 13:61-92

Brabb T, von Dassow P, Ordonez N, Schnabel B, Duke B, Goverman J (2000) In situ tolerance within the central nervous system As a mechanism for preventing autoimmunity. J Exp Med 192:871-880

Brandtzaeg P (1992) Humoral immune response patterns of human mucosae: induction and relation to bacterial respiratory tract infections. J Infect Dis 165 Suppl 1:167-176

Brienen JA, Snell CARD (1969) The mechanism of the lacrimal flow. Ophthalmologica 159: 223-232

Busse H, Müller KM (1977) Zur Entstehung der idiopathischen Dakryostenose. Klin Monatsbl Augenheilkd 170:627-32

Butcher EC, Picker LJ (1996) Lymphocyte homing and homeostasis. Science 272:60-66.

Buthler JM, Ruskell GL, Cole DF, Unger WG, Zhang SQ, Blank MA, McGregor GP, Bloom SR (1984) Effects of VIIth (facial) nerve degeneration on vasoactive intestinal polypeptide and substance P levels in ocular and orbital tissues of the rabbit. Exp Eye Res 39:523-532

Cahill KV, Burns JA (1993) Management of acute dacryocystitis in adults. Ophthal Plast Reconstr Surg 9:38-41

Calafat J, Janssen H, Tool A, Dentener MA, Knol EF, Rosenberg HF, Egesten A (1998) The bactericidal/permeability-increasing protein (BPI) is present in specific granules of human eosinophils. Blood 91:4770-4775

Cattoretti G, Pileri S, Parravicini C, Becker MH, Poggi S, Bifulco C, Key G, D'Amato L, Sabattini E, Feudale E (1993) Antigen unmasking on formalin-fixed, paraffin-embedded tissue sections. J Pathol 173:83-98

Cauna N (1970a) Electron microscopy of the nasal vascular bed and its nerve supply. Ann Oto Rhino Laryngo 79:443-450

Cauna N (1970b) The fine structure of the arteriovenous anastomosis and its nerve supply in the human nasal respiratory mucosa. Anat Rec 168:9-22

Cauna N, Cauna D (1975) The fine structure and innervation of the cushion veins of the human nasal respiratory mucosa. Anat Rec 181:1-16

Chadwick MP, Westley BR, May FEB (1997) Homodimerization and hetero-oligomerization of the single-domain trefoil protein pNR-2/pS2 through cysteine 58. Biochem J 327:117-123

Chavis RM, Welham RAN, Maisey MN (1978) Quantitative lacrimal scintillography. Arch Ophthalmol 96:2066-2068

Chen HB, Yamabayashi S, Ou B, Tanaka Y, Ohno S, Tsukahara S (1997) Structure and composition of rat precorneal tear film. A study by an in vivo cryofixation. Invest Ophthalmol Vis Sci 38:381-387

Chen Y-H, Lu Y, De Plaen IG, Wang L-Y, Tan X-D (2000) Transcription factor NF-κB signals anti-anoikic function of trefoil factor 3 on intestinal epithelial cells. Biochem Biophys Res Commun 274:576-582

Chertov O, Yang D, Zack Howard OM, Oppenheim JJ (2000) Leukocyte granule proteins mobilize innate host defenses and adaptive immune responses. Immun Rev 177:68-78

Chinery R, Bates PA, De A, Freemont PS (1995) Characterisation of the single copy trefoil peptides intestinal trefoil factor and pS2 and their ability to form covalent dimers. FEBS Lett 357:50-54

Chodosh J, Nordquist RE, Kennedy RC (1998) Anatomy of mammalian conjunctival lymphoepithelium. In Sullivan DA, Dartt DA, Stern ME, Tsubota K (eds) Second International Conference on the Lacrimal Gland, Tear Film and Dry Eye Syndromes: Basic Science and Clinical Relevance. Bermuda. Kluver Academic/Plenum Publishers, New York pp 557-565

Corfield AP, Carrington SD, Hicks SJ, Berry M, Ellingham RB (1997) Ocular mucins: purification, metabolism and functions. Prog Retin Eye Res 16:627-656
Corfield T (1992) Bacterial sialidases – roles in pathogenicity and nutrition. Glycobiology 2:509-521
Cripps MM, Bennett DJ (1992) Proenkephalin a derivate in lacrimal gland: occurrence and regulation of lacrimal function. Exp Eye Res 54:829-834
Cripps MM, Patchen-Moor K (1989) Inhibition of stimulated lacrimal secretion by met-enkephalinamide. Am J Physiol [Suppl 20] 257:G151-G156
Cullor JS, Mannis MJ, Murphy CJ, Smith WL, Selsted ME, Reid TW (1990) In vitro antimicrobial activity of defensins against ocular pathogens. Arch Ophthalmol 108:861-864
Dalgleish R (1967) Idiopathic acquired lacrimal drainage obstruction. Br J Ophthalmol 51:463-468
Danjo Y, Watanabe H, Tisdale AS, George M, Tsumura T, Abelson MB, Gipson IK (1998) Alterations of mucin in human conjunctival epithelia in dry eye. Invest Ophthalmol Vis Sci 39:2602-2609
Dartt DA (1989) Signal transduction and control of lacrimal gland protein secretion: a review. Cur Eye Res 8:619-636
Dartt DA, Baker AK, Vaillant C, Rose PE (1984) Vasoactive intestinal polypeptide stimulation of protein secretion from rat lacrimal gland acini. Am J Physiol 24: 502-509
Delventhal S, Brandis A, Ostertag H, Pabst R (1992a) Low incidence of bronchus-associated lymphoid tissue (BALT) in chronically inflamed human lungs. Virch Arch B Cell Pathol Incl Mol Pathol 62:271-274
Delventhal S, Hensel A, Petzoldt K, Pabst R (1992b) Effects of microbial stimulation on the number, size and activity of bronchus-associated lymphoid tissue (BALT) structures in the pig. Int J Exp Pathol; 73:351-357
Denton DA, Nelson JF, Tarjan E (1985) Water and salt intake of wild rabbits (Oryctolagus cuniculus (L)) following dipsogenic stimuli. J Physiol 326:285-301
Diamond G, Bevins CL (1998) Molecule of the month β-defensins: endogenous antibiotics of the innate host defense response. Clin Immunol Immunopathol 88:221-225
Dignass A, Lynch-Devaney K, Kindon H, Thim L, Podolsky DK (1994) Trefoil peptides promote epithelial migration though a transfroming growth factor β-independent pathway. J Clin Invest 94:376-383
Dilly PN (1994) Structure and function of the tear film. Adv Exp Med Biol 350:239-47
Doane MG (1981) Blinking and the mechanics of the lacrimal drainage system. Ophthalmology 88:844
Dorello U (1957) Occio e seni. Rapporti tra le affezioni dei seni e quelle vie lacrimali. Riv Otoneuro-oftal 32:613-616
Dua HS, Gomes JAP, Jindal VK, Appa SN, Schwarting R, Eagle RC, Donoso LA, Laibson PR (1994) Mucosa specific lymphocytes in the human conjunctiva, corneoscleral limbus and lcrimal gland. Curr Eye Res 13:87-93
Duke-Elder S (1961) The anatomy of the visual system: the lacrimal apparatus. System of ophthalmology Vol II. London: Kimpton pp 559-581
Duke-Elder S (1974) The ocular adnexa. System of Ophthalmology. Vol 13. London: Kimpton pp 675-770
Duke-Elder S (1952) The ocular adnexa. Textbook of Ophthalmology. Vol 5. London: Kimpton p 5302
Emami S, Le Floch N, Bruyneel E, Thim L, May F, Westley B, Rio MC, Mareel M, Gespach C (2001) Induction of scattering and cellular invasion by trefoil peptides in src- and RhoA-transformed kidney and colonic epithelial cells. FASEB J 15:351-361
Emery JL, Dinsdale F (1973) The postnatal development of lymphoreticular aggregates and lymph nodes in infants' lungs. J Clin Pathol 26:539-545
Fahrenkrug J (1993) Transmitter role of vasoactive intestinal peptide. Pharmacol Toxicol 72:354-363
Faillard H, Schauer R (1972) Glycoproteins as lubricants, protective agents, carriers, structural proteins and as participants in other functions. In: Gottschalk A (ed) Glycoproteins. Their composition, structure, and function, 2nd edn, part B, Elsevier, Amsterdam, pp 1246-1267

Feindt J, Becker I, Blömer U, Hugo H-H, Mehdorn HM, Krisch B, Mentlein R (1995) Expression of somatostatin receptor subtypes in cultured astrocytes and gliomas. J Neurochem 65:1997-2005

Fernandez-Valencia R, Gomez Pellico (1990) L Functional anatomy of the human saccus lacrimalis. Acta Anat 139:54-49

Florey HW (1933) Observations on the functions of mucus and the early stages of bacterial invasion of the intestinal mucosa. J Pathol Baceriol 37:283-289

Forstner JF (1978) Intestinal mucins in health and disease. Digestion 17: 234-263

Frieberg T (1917) Über die Mechanik der Tränenableitung. Zschr Augenheilk 37:42-66, 211-243, 324-366

Fujita R (1977) Concept of paraneurons. Arch Histol Jap Suppl 40: 1-12

Ganz T, Lehrer RI (1995) Defensins. Pharmacol Ther 66:191-205

Ganz T, Selsted ME, Szklarek D, Harwig SS, Daher K, Bainton DF, Lehrer RI (1985) Defensins: natural peptide antibiotics of human neutrophils. J Clin Invest. 76:1427-1435

Gao J, Schwab TA, Addeo JV, Ghosn CR, Stern ME (1998) The role of apoptosis in the pathogenesis of canine keratoconjunctivitis sicca: the effect of topical Cyclosporin A therapy. Cornea 17:654-663

Garfin SW (1942) Etiology of dacryocystitis and epiphora. Arch Ophthalmol 27:167-88

Gebert A, Rothkötter HJ, Pabst R (1996) M cells in Peyer's patches of the intestine. Int Rev Cytol 167:91-159

Gebert G (1995) Modes of absorption. In: Gardner MLG, Steffens K-H, eds. Absorption of orally administered enzymes. Berlin, Heidelberg, New York: Springer pp 23-28

Gendler SJ, Spicer AP (1995) Epithelial mucin genes. Annu Rev Physiol 57:607-34

Gibbons RJ (1977) Adherence of bacteria to host tissues. In: Schlessinger D (ed) Microbiology. American Society for Microbiology, Washington, DC, pp 395-406

Gibbons RJ, Houte v J (1980) Bacterial adherence and the formation of dental plaques. In: Beachey EH (ed) Receptors and recognition. Series B, vol 6, Bacterial adherence. Wiley, New York, pp 61-104

Gipson IK, Inatomi T (1997) Mucin genes expressed by the ocular surface epithelium. Prog Retin Eye Res 16:81-98

Göke MN, Cook JR, Kunert KS, Fini ME, Gipson IK, Podolsky DK (2001) Trefoil peptides promote restitution of wounded corneal epithelial cells. Exp Cell Res 264:337-344

Gormezano I, Schneiderman N, Deaux E, Fuentes I (1962) Nictitating membrane: classical conditioning and extinction in the albino rabbit. Science 138:33-34

Gorudko IV, Timoshenko AV (2000) Effect of signaling inhibitors on the release of lysozyme from human neutrophils activated by sambucus nigra agglutinin. Biochemistry 65:940-945

Gottsch JD, Li Q, Ashraf MF, Brien TPO, Stark WJ, Liu SH (1998) Defensin gene expression in the cornea. Curr Eye Res 17:1082-1086

Gould SJ, Isaacson PG (1993) Bronchus-associated lymphoid tissue (BALT) in human fetal and infant lung. J Pathol 169:229-34

Griffith TS, Brunner T, Fletcher SM, Green DR, Ferguson TA (1995) Fas ligand-induced apoptosis as a mechanism of immune privilege. Science 270:1189-1192

Groessl SA, Sires BS, Lemke Bn (1997) An anatomical basis for primary acquired nasolacrimal duct obstruction. Arch Ophthalmol 115:71-74

Haimoto H, Hosoda S, Kato K (1987) Differential distribution of immunoreactive S100α and S100β proteins in normal nonnervous tissues. Lab Invest 57:489-498

Halben R (1904) Beiträge zur Anatomie der Tränenwege. Graefe's Arch Clin Exp Ophthalmol 57:61-92

Hancock REW, Lehrer R (1998) Cationic peptides: a new source of antibiotics. Trends Biotechnol 16:8585-8589

Harder J, Bartels J, Christophers E, Schröder JM (1997) A peptide antibiotic from human skin. Nature 387:861

Harder J, Meyer-Hoffert U, Teran LM, Schwichtenberg L, Bartels J, Maune S, Schröder JM (2000) Mucoid Pseudomonas aeruginosa, TNF-alpha, and IL-1-beta, but not IL-6, induce human beta-defensin-2 in respiratory epithelia. Am J Respir Cell Mol Biol 22:714-721

Hauser F, Poulsom R, Chinery R, Rogers LA, Hanby AM, Wright NA, Hoffmann W (1993) hP1.B, a human P-domain peptide homologous with rat intestinal trefoil factor, is espressed also in the ulcer-associated cell lineage and the uterus. Proc Natl Acad Sci USA 90:6961-6965

Haynes RJ, McElveen JE, Dua HS, Tighe PJ, Liversidge J (2000) Expression of human beta-defensins in intraocular tissues. Invest Ophthalmol Vis Sci 41:3026-3031

Haynes RJ, Tighe PJ, Dua HS (1999) Antimicrobial defensin peptides of the human ocular surface. Br J Ophthalmol83:737-741

Haynes RJ, Tighe PJ, Dua HS (1998) Innate defence of the eye by antimicrobial defensin peptides. Lancet 352:451-452

Hebert E (2000) Endogenous lectins as cell surface transducers. Bioscience Rep 20: 213-37

Heinlein H (1875)Zur makroskopischen Anatomie der Thränenröhrchen. Graefe's Arch Klin Exp Ophthalmol 21:1-16

Henle J (1866) Thränenapparat. Handbuch der Eingeweidelehre des Menschen. Braunschweig, Friedrich Viehweg Verlag, pp 705-715

Higazi AA-R, Barghouti Il, Abu-Much R (1995) Identification of an inhibitor of tissue-type plasminogen activator-mediated fibrinolysis in human neutrophils: a role for defensin. J Biol Chem 270:9472-9477

Higazi AA-R, Ganz T, Kariko K, Cines DB (1996) Defensin modulates tissue-type plasminogen activator and plasminogen binding to fibrin and endothelial cells. J Biol Chem 271:17650-17655

Hill CP, Yee J, Selsted ME, Eisenberg D (1995) Crystal structure of defensins HNP-3, an: implications from an amphibian model system. Biochem Soc Transact 23:805-810

Hill JC, Bethell W, Smirmaul HJ (1974) Lacrimal drainage–a dynamic evaluation. Part I–mechanics of tear transport.Can J Ophthalmol 9:411-416

Hoffmann W, Hauser F (1993) The P-domain or trefoil motif: a role in renewal and pathology of mucous epithelia? Trends Biochem Sci 18:239-243

Hoffmann W, Jagla W, Wiede A (2001) Molecular medicine of TFF-peptides: from gut to brain. Histol Histopathol 16:319-334

Hoffmann W, Jagla W (2001) Cell type specific expression of secretory TFF-peptides: colocalization with mucins and synthesis in the brain. Internat. Rev. Cytol. 2001b; 213:147-181.

Hoffmann W, Joba W (1995) Biosynthesis and molecular architecture of gel-forming mucins: implications from an amphibian model system. Biochem Soc Trans 23:805-810

Hurwitz JJ (1996) In: Hurwitz JJ (ed) The lacrimal system. Lippincott-Raven, Philadelphia, chapter 4, pp 23-29

Hurwitz JJ, Maisey MN, Welham RAN (1975a) Quantitative lacrimal scintillography. Method and physiological application. Br J Ophthalmol 59:308-312

Hurwitz JJ, Welham RAN, Lloyd GAS (1975b) The role of the intubation macrodacryocystography in management of problems of the lacrimal system. Can J Ophthalmol 10:361-368

Hutch JA (1970) The role of urethral mucus in the bladder defense mechanism. J Urol 103: 165-167

Hyjek E, Smith WJ, Isaacson PG (1988) Primary B cell lymphoma of salivary glands and its relationship to myoepithelial sialadenitis. Hum Pathol 19:766-776

Inatomi T, Spurr-Michaud S, Tisdale AS, Zhan Q, Feldman ST, Gipson I (1996) Expression of secretory mucin genes by human conjunctival epithelia. Invest Ophthaomol Vis Sci 37:1684-1692

Isaacson PG (1992) Extranodal lymphomas: The MALT concept. Verh Dtsch Ges Pathol 76:14-23

Islam D, Bandholtz L, Nilsson J, Wigzell H, Christensson B, Agerberth B, Gudmundsson GH (2001) Downregulation of bactericidal peptides in enteric infections: A novel immune escape mechanism with bacterial DNA as a potential regulator. Nat Med 7:180-185

Iwata N (1927) Beiträge zur Kenntnis der Formverhältnisse der Tränenwege des Menschen mit besonderer Berücksichtigung ihrer Entwicklung. Folia Anat Japn 5: 51-168

Jagla W, Wiede A, Dietzmann K, Rutkowski K, Hoffmann W (2000) Co-localization of TFF3 peptide and oxytocin in the human hypothalamus. FASEB J 14:1126-1131

Jagla W, Wiede A, Hinz M, Dietzmann K, Gulicher D, Gerlach KL, Hoffmann W (1999a) Secretion of TFF-peptides by human salivary glands. Cell Tissue Res. 298:161-166

Jagla W, Wiede A, Hoffmann W (1999b) Localization of TFF3 peptide to porcine conjunctival goblet cells. Cell Tissue Res 296:525-530

Jagla W, Wiede A, Kölle S, Hoffmann W (1998) Differential expression of the TFF-peptides xP1 and xP4 in the gastrointestinal tract of Xenopus laevis. Cell Tissue Res 291:13-18

Janssen AG, Mansour K, Bos JJ, Castelijns JA (2001) Diameter of the bony lacrimal canal: normal values and values related to nasolacrimal duct obstruction: assessment with CT. AJNR Am J Neuroradiol 22:845-850

Janssen PT, van Bijsterveld OP (1983) The relations between tear fluid concentrations of lysozyme, tear-specific prealbumin and lactoferrin. Exp Eye Res 36:773-779

Joers K (1899) Beiträge zur normalen und pathologischen Histologie des Tränenschlauches. Beitr Augenheilk 4:355-398

Johannson O, Lundberg JM (1981) Ultrastructural localization of VIP-like immunoreactivity in large dense core vesicles of „cholinergic-type" nerve terminals in cat exocrine glands. Neuroscience 6:847-862

Jolly CL, Beisner BM, Holmes IH (2000) Rotavirus infection of MA104 cells is inhibited by ricinus lectin and separately expressed single binding domains. Virology 275:89-97

Jones DT, Ji A, Monroy D, Ji Z, Pflugfelder SC (1998) Alterations of ocular surface gene expression in Sjogren's syndrome. Adv Exp Med Biol 438:533-536

Jones GW (1977) The attachment of bacteria to the surface of animal cells. In: Ressig JL (ed) Receptors and recognition. Series B, vol 3. Microbiol interactions. Wiley, New York, pp 139-176

Jones LT (1961) An anatomical approach to problems of the eyelids and lacrimal apparatus. Arch Ophthalmol 66: 111-124

Jones LT (1958) Practical fundamental anatomy and physiology of the eye. Trans Am Acad Ophthalmol Otolaryngol 62: 669-678

Jones LT, Boyden GL (1972) The lacrimal apparatus. In: Maloney W, ed. Otolaryngology. Hagerstown: Haper & Row, Volume 3, Chapter 29

Jones LT (1957) Epiphora II. Its relation of the anatomic structures and surgery of the medial cantal region. Am J Ophthalmol 43:203-212

Jumblatt MM, McKenzie RW, Jumblatt JE (1999) MUC5AC mucin is a component of the human precorneal tear film. Invest Ophthalmol Vis Sci 40:43-49

Kagan BL, Ganz T, Lehrer RI (1994) Defensins: a family of antimicrobial and cytotoxic peptides. Toxicology 87:131-149

Kahn J, Anderton BH, Miller CC, Wood JN, Esiri MM (1987) Staining with monoclonal antibodies to neurofilaments distinguishes between subpopulations of neurofibrillary tangles, between groups of axons and between groups of dendrites. J Neurol 234:241-246

Kelleher RS, Hann LE, Edwards JA, Sullivan DA (1991) Endocrine, neural, and immune control of secretory component output by lacrimal gland acinar cells. J Immunol 146: 3405-3412

Kelm S, Schauer R (1997) Sialic acids in molecular and cellular interactions. Int Rev Cytol 175: 137-240

Kheterpal S, Chan SY, Batch A, Kirkby GR (1994) Previously undiagnosed lymphoma presenting as recurrent dacryocystitis. Arch Ophthalmol 112:519-20

Kijlstra A (1990) The role of lactoferrin in the nonspecific immune response on the ocular surface. Regul Immunol. 3:193-197

Kilian M, Mestecky J, Russell MW (1988) Defense mechanisms involving Fc-dependent functions of immunglobulin A and their subversion by bacterial immunglobulin A proteases. Microbiol Rev 52: 296-303

Kindon H, Pothoulakis C, Thim L, Lynch-Devaney K, Podolsky DK (1995) Trefoil peptide protection of intestinal epithelial barrier function: cooperative interaction with mucin glycoprotein. Gastroenterology 109:516-23

Kinoshita T, Taupin DR, Itoh H, Podolsky DK (2000) Distinct pathways of cell migration and antiapoptotic response to epithelial injury: structure-function analysis of human intestinal trefoil factor. Mol Cell Biol 20:4680-4690

Knipping S, Riederer A, Fischer A (1995) Immunhistochemische Untersuchungen zur Neuroanatomie der menschlichen Nasenmuschel: Innervationsmuster seröser Drüsen. Laryngo Rhino Oto 74: 81-84

Knop E, Knop N (2001) Lacrimal drainage-associated lymphoid tissue (LDALT): a part of the human mucosal immune system. Invest Ophthalmol Vis Sci 42:566-574
Knop N, Knop E (2000) Conjunctiva-associated lymphoid tissue in the human eye. Invest Ophthalmol Vis Sci 41:1270-1279
Köhler U, Müller W (1968) Nasennebenhöhlenbefunde bei röntgenologisch gesicherten Dacryostenosen. Klin Monatsbl Augenheilkd 152:813-816
Körner F (1937) Über Drosselvenen im Schwellgewebe der Nasenschleimhaut. Z mikrosk-anat Forsch 41: 131-150
Krehbiel GAA (1878) Die Muskulatur der Thränenwege und der Augenlider mit spezieller Berücksichtigung der Thränenleitung. Cotta'sche Buchhandlung, Stuttgart, pp 1-36
Kudryashov BA, Kondashevskay MV, Lyapina LA (1990) Action of defensin on healing of aseptic skin wounds and vascular permeability. Bull Exp Biol Med 109:513-515
Kuper CF, Loornstra PJ, Hameleers DMH, Biewenga J, Spit BJ, Duijvestijn AM, van Breda Vriesman PJ, Sminia T (1992) The role of nasopharyngeal lymphoid tissue. Immunol Today 13:219-224
Langer G, Jagla W, Behrens-Baumann W, Walter S, Hoffmann W (2002) Ocular TFF-peptides: new mucus-associated secretory products of conjunctival goblet cells. In: Sullivan DA (ed) Lacrimal gland, tear film, and dry eye syndroms 3. Kluwer Academic, Plenum Publishers, New York, in press
Langer G, Jagla W, Behrens-Baumann W, Walter S, Hoffmann W (1999) Secretory peptides TFF1 and TFF3 synthesized in human conjunctival goblet cells. Invest Ophthalmol Vis Sci 40:2220-2224
Langer G, Walter S, Behrens-Baumann W, Hoffmann W (2001) TFF-Peptide: neue Mukus-assoziierte Sekretionsprodukte der Konjunktiva. Ophthalmologe 98:976-979.
Lefebvre O, Wolf C, Kedinger M, Chenard MP, Tomasetto C, Chambon P, Rio MC (1993) The mouse one P-domain (pS2) and two P-domain (mSP) genes exhibit distinct patterns of expression. J Cell Biol 122:191-198
Lehmann J (1990) Neuropeptide Y: an overview. Drug Dev Res 19:329-351
Lehmann OJ, Hussain IR, Watt PJ (2000) Investigation of β defensin gene expression in the ocular anterior segment by semiquantitative RT-PCR. Br J Ophthalmol 84:523-526
Lehrer RI, Barton A, Daher KA, Harwig SS, Ganz T, Selsted ME. Interaction of human defensins with Escherichia coli. Mechanism of bactericidal activity. J Clin Invest. 1989; 84:553-561.
Lehrer RI, Ganz T, Selsted ME. Defensins: endogenous antibiotic peptides of animal cells. Cell. 1991;64:229-230.
Lemp MA (1995) Report of the National Eye Institute/Industry Workshop on Clinical Trials in Dry Eyes. CLAO J 21:222-232
Lemp MA (1973) Surfacing the precorneal tear film. Ann Ophthalmol 5:819-26
Leone CR (1986) Nasolacrimal duct obstruction. Ophthalmology 93:1063
Leone GBC (1574) In altero de musculis palpebrarum atque oculorum motibus deservientibus, accurate disseritur. Anatomici libri II, Ticini
Lesshaft P (1868) Über den Musculus orbicularis orbitae und seinen Einfluß auf den Mechanismus der Tränenabsonderung. Arch Anat Physiol pp 265-301
Linberg JV, McCormick SA (1986) Primary acquired nasolacrimal duct obstruction. A clinical pathologic report and biopsy technique. Ophthalmology 93:1055-1063
Lindvall UO, Björklund A, Falk B (1981) Fluorescence microscopy of biogenic amines. In: Methods in neurobiology 2. Lahne R (ed) New York, London, Plenum Press, pp 365-431
Linehan SA, Martinez-Pomares L, Gordon S (2000) Macrophage lectins in host defence. Microbes Infect 2:279-88
Lundberg JM, Ånggård A, Fahrenkrug J, Hökfeld T, Mutt V (1980) Vasoactive intestinal polypeptide in cholinergic neurons of exocrine glands: functional significance of coexisting transmitters for vasodilation and secretion. Proc Nat Acad Sci USA 77:1651-1655
Lundberg JM, Martling CR, Lundblad L (1988) Cigarette smoke-induced irritation in the airways in relation to peptide-containing, capsaicin-sensitive sensory neurons. Klin Wochenschr 66 Suppl 11:151-60

Lundberg JM, Terenius L, Hökfeld T, Martling CR, Tatemoto K, Mutt V, Polak J, Bloom S, Goldstein M (1982) Neuropeptide Y (NPY)-like immunoreactivity in peripheral noradrenergic neurons and effects of NPY on sympathetic function. Acta Physiol Scand 116: 477-480

Macdonald EA, Maurice DM (1991) Loss of fluorescein across the conjunctiva. Exp Eye Res 53:427-430

Maltzman BA, Favetta JR (1979) Dacryolithiasis. Ann Ophthalmol 11:473-475

Markovitch A (1951) Considération sur la division de l'appareil lacrymal. Ann Oculist 184:803-810

Marra MN, Wilde CG, Griffith JL, Snable M, Thornton B, Scott RW (1992) The role of bactericidal/permeability-increasing protein as a natural inhibitor of bacterial endotoxin. J Immunol148:532-537

Mashimo H, Wu DC, Podolsky DK, Fishman MC (1996) Impaired defense of intestinal mucosa in mice lacking intestinal trefoil factor. Science 274:262-265

Matsumoto Y, Tanabe T, Ueda S, Kawata M (1992) Immunohistochemical and enzymehistochemical studies of peptidergic, aminergic and cholinergic innervation of the lacrimal gland of the monkey (Macaca fusata). J Aut Nerv Sys 37:207-214

Maurice DM (1952) ce on corneal permeability of bathing with solutions of differing reaction on tonicity. Br J Ophthalmol 39:463-473

Maurice DM (1973) The dynamics and drainage of tears. Int Ophthalmol Clin 13:103-116

Maurice DM (1995) The effect of the low blink rate in rabbits on topical drug penetration. J Ocul Pharmacol Ther 11:279-304

Maurice DM (1967) The use of fluorescein in ophthalmological research. Invest Ophthalmol Vis Sci 6:464-477

Mauriello JA, Palydowycz S, DeLuca J (1992) Clinicopathologic study of lacrimal sac and nasal mucosa in 44 patients with complete acquired nasolacrimal duct obstruction. Ophthalmic Plast Reconstr Surg 8:13-21

Mc Carty CA, Bansal AK, Livingston PM, Stanislavsky YL, Taylor HR (1998) The epidemiology of dry eye in Melbourne, Australia. Ophthalmology 105:1114-1119

McEwen WK (1962) Secretion of tears and blinking. In: Davson H (ed) The eye. Vol 3. Academic Press, New York, pp 271-301

McNamara NA, Van R, Tuchin OS, Fleiszig SMJ (1999) Ocular surface epithelia express mRNA for human beta defensin-2. Exp Eye Res 69:483-490

Medawar PB (1948) Immunity to homologous grafted skin. III. The fate of skin homografts transplanted to the brain, to subcutaneous tissue, and to the anterior chamber of the eye. British J Exp Pathol 29: 58-69

Melanová J (1969) Diverticulum of lacrimal sac. Čsl Oftal 47:47-49

Meller J (1929) Diseases of the lacrimal apparatus. Trans Ophthal Soc UK 49:233-311

Mestecky J, McGhee JR (1987) Immunglobulin A (IgA): molecular and cellular interactions involved in IgA biosynthesis and immune response. Adv Immunol 40:153-245

Milder B (1983) Clinical biochemistry of tears. Composition of tear fluid. In: Milder B (ed) The lacrimal system. Appleton-Centrury-Crofts, Conneticut, pp 35-48

Moore BW (1975) Brain specific protein: S-100 protein, 14-3-2 protein, and glial fibrillary protein. Adv Neurochem 1:137

Moore CP, Wilsman NJ, Nordheim EV, Majors LJ, Collier LL (1987) Density and distribution of canine conjunctival goblet cells. Invest Ophthalmol Vis Sci 28:1925-1932

Müller W, Köhler U, Rose KG (1968) Die Bedeutung der Sinusitis maxillaris für die Entwicklung der Dacryostenose. Klin Monatsbl Augenheilkd 153:660-662

Murphy CJ, Foster BA, Mannis MJ, Selsted ME, Reid TW (1993) Defensins are mitogenic for epithelial cells and fibroblasts. J Cell Physiol 155:408-413

Murube del Castillo J (1978) On gravity as one of the impelling forces of lacrimal flow. Asahi Evening News. January 20, pp 51-59

Nagashima K, Araki K (1963) On the lacrimal part of the orbicularis oculi muscle with special reference to the sac dilators. Jap J Ophthalmol 7: 220-225

Nielsen PA, Bennett EP, Wandall HH, Therkildsen MH, Hannibal J, Clausen H (1997) Identification of a major human high molecular weight salivary mucin (MG1) as tracheobronchial mucin MUC5B. Glycobiology 7:413-419

Nielsen PA, Mandel U, Therkildsen MH, Clausen H. (1996) Differential expression of human high-molecular-weight salivary mucin (MG1) and low-molecular-weight salivary mucin (MG2). J Dent Res 75:1820-1826

Nik NA, Hurwitz JJ, Chin Sang H (1984) The mechanism of tear flow after DCT and Jones' tube surgery. Arch Ophthaomol 102:1643-1649

Nitsch R, Bechmann I, Deisz RA, Haas D, Lehmann TN, Wendling U, Zipp F (2000) Human brain-cell death induced by tumour-necrosis-factor-related apoptosis-inducing ligand. Lancet 356:827-828

Oertel M, Graneß A, Thim L, Bühling F, Kalbacher H, Hoffmann W (2001) Trefoil factor family-peptides promote migration of human bronchial epithelial cells: synergistic effect with epidermal growth factor. Am J Respir Cell Mol Biol 25:418-424

Ohtani O, Ushiki T, Taguchi T, Kikuta A (1988) Collagen fibrillar networks as skeletal frameworks: A demonstration by cell maceration/scanning electron microscope method. Arch Histol Cytol 51:249-261

Oram J, Reiter B (1979) Inhibition of bacteria by lactoferrin and other iron-chelating agents. Biochim Biophys Acta 170:351-353

Österlind G (1944) An investigation into the presence of lymphatic tissue in the human conjunctiva, and its biologic and clinical importance. Acta Ophthalmol 23:1-79

Pabst R, Gehrke I (1990) Is the bronchus-associated lymphoid tissue (BALT) an integral structure of the lung in normal mammals including humans? Am J Respir Cell Mol Biol 3:131-135

Parke WW, Whalen JL, Bunger PC, Settles HE (1995) Intimal musculature of the lower anterior spinal artery. Spine 20:2073-2079

Passo MS, Palmer EA, van Buskirk EM (1984) Plasma Timolol in glaucoma patients. Ophthalmology 91:1361-1363

Paulsen F (2001) Die ableitenden Tränenwege des Menschen – Abwehrfunktion und Biomechanik. Habilitationsschrift, Universität Kiel

Paulsen F, Föge M, Thale A, Tillmann B, Mentlein R (2002e) Animal model for the absorption of lipophilic substances from tear fluid by the epithelium of the nasolacrimal ducts. Invest Ophthalmol Vis Sci (in press)

Paulsen F, Hallmann U, Paulsen J, Thale A (2000a) Innervation of the cavernous body of the human efferent tear ducts and function in tear outflow mechanism. J Anat 196:177-187

Paulsen F, Hinz M, Schaudig U, Thale A, Hoffmann W (2002a) TFF-peptides in efferent tear ducts. Invest Ophthaolmol Vis Sci (in press)

Paulsen F, Paulsen J, Thale A, Schaudig U, Tillmann B (2002b) Organized mucosa associated lymphoid tissue in human nasolacrimal ducts. In: Sullivan DA, ed. Lacrimal gland, tear film and dry eye syndromes 3. New York: Kluver Academic, Plenum Publishers (in press)

Paulsen F, Paulsen J, Thale A, Tillmann B (2000b) Mucosa-associated lymphoid tissue (MALT) in the human efferent tear ducts. Virch Arch 437:185-189

Paulsen F, Pufe T, Schaudig U, Held-Feindt J, Lehmann J, Schröder J-M, Tillmann B (2001a) Detection of natural peptide antibiotics in human nasolacrimal ducts. Invest Ophthalmol Vis Sci 42:2157-2163

Paulsen F, Pufe T, Schaudig U, Held-Feindt J, Lehmann J, Thale A, Tillmann B (2002c) Protection of human efferent tear ducts by antimicrobial peptides. In Sullivan DA (ed), Lacrimal gland, tear film and dry eye syndromes 3. Kluver Academic, Plenum Publishers, New York (in press)

Paulsen F, Thale A (1998) Epithelial-connective tissue boundary in the oral part of the human soft palate. J Anat 193:457-467

Paulsen F, Thale A, Kohla G, Schauer R, Rochels R, Parwaresch R, Tillmann B (1998) Functional anatomy of human lacrimal duct epithelium. Anat Embryol 198:1-12

Paulsen F, Thale A, Maune S, Tillmann B (2001b) Primary acquired dacryostenosis – histopathology and pathophysiology. Ophthalmology 108:2329-2336

Paulsen F, Thale A, Mentlein R (2000c) What happens to tears inside the efferent lacrimal passage? – An animal experimental study. Graefe's Arch Clin Exp Ophthalmol 238:496-499

Paulsen F, Thale A, Schaudig U (2002d) Ableitende Tränenwege und Trockenes Auge. Ophthalmologe 2002; (in press)

Paulsen FP, Thale AB, Hallmann UJ, Schaudig U, Tillmann BN (2000d) The cavernous body of the human efferent tear ducts: function in tear outflow mechanism. Invest Ophthalmol Vis Sci 41:965-970

Perra MT, Serra A, Sirigu P, Turno F (1995) A histochemical and immunohistochemical study of certain defense mechanisms in the human lacrimal sac epithelium. Arch Histol Cytol 58:517-522

Pflugfelder SC, Solomon A, Stern ME (2000) The diagnosis and management of dry eye. Cornea 19:644-649

Pflugfelder SC, Tseng SC, Sanabria O, Kell H, Garcia CG, Felix C, Feuer W, Reis BL (1998) Evaluaiton of subjective assessments and objective diagnostic tests for diagnosing tear-film disorders known to cause ocular irritation. Cornea 17:38-56

Pflugfelder SC, Tseng SCG, Yoshino K, Monroy D, Felix C, Reis BL (1997) Correlation of goblet cell density and mucosal epithelial membrane mucin expression with rose bengal staining in patients with ocular irritations. Ophthalmology 104:223-235

Pigott NH, Henry JA, May FEB, Westley BR (1991) Antipeptide antibodies against the pNR-2 oestogen-regulated protein of human breast cancer cells and detection of pNR-2 expression in normal tissues by immunohistochemistry. J Pathol 163:95-104

Playford RJ (1995) Peptides and gastrointestinal mucosal integrity. Gut 37:595-597

Prohaszka Z, Fust G (1998) Contribution of complement to defensin action in eye (Letter). Lancet352:1152

Prohaszka Z, Nemet K, Csermely P, Hudecz F, Mezo G, Fust G (1997) Defensins purified from human granulocytes bind C1q and activate the classical complement pathway like the transmembrane glycoprotein gp41 of HIV-1. Mol Immunol 34:809-816

Qu X-D, Lehrer RI (1998) Secretory phospholipase A_2 is the principal bactericide for staphylococci and other gram-positive bacteria in human tears. Infect Immun 66:2791-2797

Radnot M (1977) Die Flimmerhaare des Tränensackepithels. Klin Monatsbl Augenheilkd 170:428-432

Radnot M, Bölcs S (1971) Die Feinstruktur der Epithelzellenoberfläche des Tränensackes. Klin Monatsbl Augenheilkd 159:158-164

Reuter G, Schauer R, Bumm P (1988) Sialic acids of human nasal mucin, possible targets of influenza C viruses. In: Schauer R, Yamakawa T (eds) Sialic acids. Proceedings of the Japanese-German Symposium on Sialic Acids. Berbel-Mendel, Kiel, pp 259-260

Reuter G, Struwe R, Feige J, Brede R, Bumm P, Schauer R (1992) Analysis of carbohydrate composition and sialidase activity in oral secretions of patients with tumors in the upper aerodigestive tract. Eur Arch Otorhinolaryngol 249:5-11

Rheeman CH, Meyer DR (1998) Enlargement of the nasolacrimal canal in the absence of neoplasia. Ophthalmology 105:1498-1503

Riederer A, Greves G, Welsch U, Herzmann S (1997) Elektronenmikroskopische Untersuchungen zur Gefäßinnervation der Nasenschleimhaut des Menschen. Laryngo Rhino Oto 76:405-410

Rio MC, Bellocq JP, Daniel JY, Tomasetto C, Lathe R, Chenard MP, Batzenschlager A, Chambon P (1988) Breast cancer-associated pS2 protein: synthesis and secretion by normal stomach mucosa. Science 241:705-708

Rivas L, Rodriguez JJ, Murube J (1991) Glandulas serosas en el saco lagrimal. Arch Soc Esp Oftalmol 60:173-176

Robbins PD, Hahnel E, Sterrett GF, Harvey J, Carrello S, Hahnel R (1993) Expression of the pS2 gene in breast cancer–a comparison of pS2 protein radioimmunoassay and immunohistochemistry. Pathology 25:255-260

Robinson SE, Schwarz JP, Costa E (1980) Substance P in the superior cervical ganglion and the submaxillary gland of the rat. Brain Res 182:11-17

Rohen JW (1953) Morphologische Studien zur Funktion des Lidapparates beim Menschen. Morph Jahrbuch 93:42-97

Rohen JW (1964) Haut und Sinnesorgane. In: Möllendorf W von, ed. Handbuch der mikroskopischen Anatomie des Menschen. Vol. 3. Berlin: Springer pp 448-451

Romeis B (1989) Färben der Schnitte. In: Böck P (ed) Mikroskopische Technik. 17th ed. Baltimore, Urban & Schwarzenberg, pp 179-249

Rosengren B (1972) On lacrimal drainage. Ophthalmologica 164:409-421

Rosengren B (1928) Studien über die Tränenableitung beim Menschen und einigen Säugetieren. Svenska läkare-Sällskapets Handlingar 54:41-117
Rosenstock T, Hurwitz JJ (1982) Functional obstruction of the lacrimal drainage passages. Can J Ophthalmol 17:249-255
Rudich L, Butcher FR (1976) Effects of substance P and eledoisin on K efflux, amylase release and cyclic nucleotide levels in slices of rat parotid gland. Biochem Biophys Acta 444:704-711
Ruiz Barranco F (1966) Patogenia de las dacrocistitis papel del conducto nasal. Arch Soc Oftal Hisp-Amer 26:113-118
Santos-Fernandez J (1921) The measurements of the nasal canal according to the face. Am J Ophthalmol 4:32-37
Sasaki T, Tanaka N, Odagiri Y, Ito D, Takebayashi E (1973) Microbial flora in dacryocystitis. Acta Soc Ophthalmol Jpn 77 :644-648
Schaeffer JP (1922) The genesis and development of the nasolacrimal passages in man. Am J Anat 13:1-24
Schaudig U, Paulsen F, Thale A (2001) Funktionelle Anatomie und Physiologie der ableitenden Tränenwege. Z prakt Augenheilkd 22:53-62
Schauer R (1992) Sialinsäurereiche Schleime als bioaktive Schmierstoffe. Nachr Chem Tech Lab 40:1227-1231
Schauer R (1997) Enzyme als Werkzeuge für die Analyse der Struktur und Funktion von sialylierten Glykokonjugaten. BIO-forum 20:541-544
Schirmer O (1903) Studien zur Physiologie und Pathologie der Tränenabsonderung und Tränenabfuhr. Arch Ophthalmol 56:197-212
Schmechel D, Marangos SS, Brightman T (1978) Neuron-specific enolase is a molecular marker for peripheral and central neuroendocrine cells. Nature 276: 834-836
Schnyder F (1920) Über familiäres Vorkommen resp. die Vererbung von Erkrankungen der Tränenwege. Z Augenheilk 44:257-69
Schröder JM (1999) Epithelial antimicrobial peptides: innate local host response elements. Cell Mol Life Sci 56:32-46
Schröder JM, Harder J (1999) Human beta-defensin-2. Int J Biochem Cell Biol 31:645-651
Schwab IR, Dries D, Cullor J, Smith W, Mannis M, Reid T, Murphy CJ (1992) Corneal storage medium preservation with defensins. Cornea 11:370-375
Schwarz H, Jagla W, Wiede A, Hoffmann W (2001) Ultrastructural co-localization of TFF3-peptide and oxytocin in the neural lobe of the porcine pituitary. Cell Tissue Res 305:411-416
Seifert P, Spitznas M (1994) Demonstration of nerve fibres in human accessory lacrimal glands. Graefés Arch Clin Exp Ophthalmol 232:107-114
Seifert P, Spitznas M (1996) mmunocytochemical and ultrastructural evaluation of the distribution of nervous tissue and neuropeptides in the Meibomian gland. Graefés Arch Clin Exp Ophthalmol 234:648-656
Seifert P, Spitznas M, Koch F, Cusumano A (1993) he architecture of human accessory lacrimal glands. German J Ophthalmol 2:444-454
Seifert P, Stuppi S, Spitznas M (1997) istribution pattern of nervous tissue and peptidergic nerve fibres in accessory lacrimal glands. Cur Eye Res 16:298-302
Seifert P, Stuppi S, Spitznas M, Weihe E (1996) Differential distribution of neuronal markers and neuropeptides in the human lacrimal gland. Graefe's Arch Clin Exp Ophthalmol 234:232-240
Shafer WM, Martin LE, Spitznagel JK (1986) Late intraphagosomal hydrogen ion concentration favors the in vitro antimicrobial capacity of a 37-kilodalton cationic granule protein of human neutrophil granules. Infect Immun53:651-655
Sharon N, Lis H (1972) Lectins: cell-agglutinating and sugar-specific proteins. Science 177:949-959
Shirai Y (1921) Jap Med World 1921;1:14
Sibony PA, Walcott B, McKeon C, Jakobiec FA (1988) Vasoactive intestinal polypeptide and the innervation of the human lacrimal gland. Arch Ophthalmol 106:1085-1088
Singh J, Adeghate E, Burrows S, Howarth F, Donath T (1994) Protein secretion and the identification of neurotransmitters in the isolated pig lacrimal gland. In: Lacrimal gland, tear film, and dry eye syndromes: basic science and clinical relevance. Sullivan DA (ed) New York, Plenum Press, pp 57-60

Singh PK, Jia HP, Wiles K, Hesselberth J, Liu L, Conway BA, Greenberg EP, Valore EV, Welsh MJ, Ganz T, Tack BF, McCray PB (1998) Production of beta-defensins by human airway epithelia. Proc Natl Acad Sci USA 95:14961-14966
Singh PK, Tack BF, McCray PB, Welsh MJ (2000) Synergistic and additive killing by antimicrobial factors found in human airway surface liquid. Am J Physiol Lung Cell Mol Physiol 279:L799-805
Sirigu P, Maxia C, Puxeddu R, Zucca I, Piras F, Perra MT (2000) The presence of a local immune system in the upper blind and lower part of the human nasolacrimal duct. Arch Histol Cytol 63:431-439.
Sood NN, Ratnaraj A, Balaraman G, Madhavan HN (1967) Chronic dacryocystitis; a clinical-bacteriological study. J All-India Ophthalmol Soc 15:107-10
Spaeth GL (1967) Nasolacrimal duct obstruction caused by topical epinephrine. Arch Ophthalmol 77:355-357
Stanculéanu G (1900) Recherches sur le dévelopement des voies lacrymalis chez l'homme et chez les animaux. Arch Ophthalmol 20:141-194
Steinkogler FJ (1986) The postsaccal, idiopathic dacryostenosis – experimental and clinical aspects. Doc Ophthalmol 63:265-286
Stensen N (1662) De ductu salivali novo, saliva, ductibus oculorum aquosis et humore oculi aqueo. Quibus varia oris oculorum et narium vasa discributu, novique salivae lacrymae et muci fontis deteguntur et novum norilissimi Bilsii de lymphae notu et usu comentum examinatur et rehictur. Lugdune Baravorum. Observationes anatomicae, Leyden
Stern ME, Beuerman RW, Fox RI, Gao J, Mircheff AK, Pflugfelder SC (1998) The pathology of dry eye: the interaction between the ocular surface and lacirmal glands. Cornea 17:584-589
Stieve H (1928) Ueber die Bedeutung venöser Wundernetze für den Verschluß einzelner Oeffnungen des menschlichen Körpers. Dtsch Med Wschr 54:87-90, 130-134
Streilein JW, Niederkorn JY (1985) Characterization of the suppressor cell(s) responsible for anterior chamber-associated immune deviation (ACAID) induced in BALB/c mice by P815 cells. J Immunol 134:1381-1387
Streilein JW (1996) Peripheral tolerance induction: lessons from immune privileged sites and tissues. Transplant Proc 28:2066-2070
Sullivan DA, Krenzer KL, Sullivan BD (1999) Does androgen insufficiency cause lacrimal gland inflammation and aqueous tear deficiency? Invest Ophthalmol Vis Sci 40:1261-1265
Summerskill WH (1956) Problems of lacrimal obstruction The rhinological approach. Trans Ophthalmol Soc UK 1956; 76:385-96.
Tarbet KJ, Custer PL (1995) External dacryocystorhinostomy: surgical success, patient satisfaction, and economic cost. Ophthalmology 102:1065-1070
Tarjan E, Denton DA, McKinley MJ, Nelson JF, Weisinger RS. What makes wild rabbits drink? J Physiol. (Paris) 1984; 79:466-470.
Taupin D, Ooi K, Yeomans N, Giraud A (1996) Conserved expression of intestinal trefoil factor in the human colonic adenoma-carcinoma sequence. Lab Invest 75:25-32
Taupin D, Pedersen J, Familari M, Cook G, Yeomans N, Giraud AS (2001) Augmented intestinal trefoil factor (TFF3) and loss of pS2 (TFF1) expression precedes metaplastic differentiation of gastric epithelium. Lab Invest 81:397-408
Taupin DR, Kinoshita K, Podolsky DK (2000) Intestinal trefoil factor confers colonic epithelial resistance to apoptosis. Proc Natl Acad Sci USA 97:799-804
Temesrékási D (1969) Mikroskopischer Bau und Funktion des Schwellgewebes der Nasenmuschel des Menschen. Z mikrosk-anat Forsch 80:219-229
Territo MC, Ganz T, Selsted ME, Lehrer RI (1989) Monocyte-chemotactic activity of defensins from human neutrophils. J Clin Invest 84:2017-2020
Tervo K, Tervo T, Eränkö O, Valtonen S, Cuello C (1982) Effect of sensory and sympathetic denervation on substance P immunoreactivity in nerve fibres of the rabbit eye. Exp Eye Res 34:577-585
Thale A, Paulsen F, Kohla G, Schauer R, Rochels R, Tillmann B (2001) Die ableitenden Tränenwege unter physiologischen und immunologischen Gesichtspunkten. Ophthalmologe 98:56-63
Thale A, Paulsen F, Rochels R, Tillmann B (1997) Functional anatomy of human efferent tear ducts. A histological, immunohistochemical and scanning electron microscopical study. In:

McCafferty G, Coman W, Carroll R (eds) XVI World Congress of Otorhinolaryngology Head and Neck Surgery, Sydney, Monduzzi Editore, Bologna, pp 1587-1590
Thale A, Paulsen F, Rochels R, Tillmann B (1998) Functional anatomy of the human efferent tear ducts: a new theory of tear outflow mechanism. Graefe's Arch Clin Exp Ophthalmol 236:674-678
Tomasetto C, Masson R, Linares JL, Wendling C, Lefebvre O, Chenard MP, Rio MC (2000) pS2/TFF1 interacts directly with the VWFC cysteine-rich domains of mucins. Gastroenterology. 118:70-80
Traquair HM (1941) Chronic dacryocystitis; its causation and treatment. Arch Ophthalmol 26:165-180
Tschernig T, Kleemann WJ, Pabst R (1995) Bronchus-associated lymphoid tissue (BALT) in the lungs of children who had died from sudden infant death syndrome and other causes. Thorax 50:658-660
Tsuda K (1952) On histology of ductus lacrimalis in adult, especially on its innervation. Tohoku J Esp Med 56:233-243
Tsukahara S, Tanishima T (1974) Adrenergic and cholinergic innervation of the human gland. Jap J Ophthalmol 18:70-77
Tucker N, Chow D, Stockl F, Codere F, Burnier M (1997) Clinically suspected primary acquired nasolacrimal duct obstruction. Ophthalmology 104:1882-1886
Tucker NA, Codere F (1994) The effect of fluorescein volume on lacrimal outflow transit time. Ophthal Plast Reconstr Surg 10:256-259
Tzou P, Ohresser S, Ferrandon D, Capovilla M, Reichhart J, Lemaitre B, Hoffmann JA, Imler J (2000) Tissue-specific inducible expression of antimicrobial peptide genes in drosophila surface epithelia. Immunity 13:737-748
Veh RW, Corfield AP, Schauer R, Andres KH (1979) The bovine submandibular gland. 2. Histochemical and biomechanical results. In: Schauer R, Boer P, Guddecke E, Kramer MF, Vliegenthart JFG, Wiegandt H (eds) Glycoconjugates. Thieme, Stuttgart, pp 193-194
Verdugo P (1990) Goblet cells secretion and mucogenesis. Annu Rev Physiol 52:157-176
Walker WA (1976) Host defense mechanisms in the gastrointestinal tract. Pediatrics 57:901-916
Weersink AJL, van Kessel KPM, van den Tol ME, can Strijp AG, Torensma R, Verhoef J, Elsbach P, Weiss J (1993) Human granulocytes express a 55-kDa lipopolysaccharide binding protein on the cell surface that is identical to the bactericidal/permeability-increasing protein. J Immunol 150:253-263
Wenkel H, Streilein JW, Young MJ (2000) Systemic immune deviation in the brain that does not depend on the integrity of the blood-brain barrier. J Immunol 164:5125-5131
Werncke T (1905) Ein Beitrag zur Anatomie des Tränensackes, speziell zur Frage der Tränensackdrüsen. Klin Monatsbl Augenheilkd 43: 191-205
West JM (1909) A window resection of the nasal duct in case of stenosis. Trans Am Ophthalmol Soc 12:659
White ML, Ma JK, Birr CA, Trown PW, Carroll SF (1994) Measurement of bactericidal/permeability-increasing protein in human body fluids by sandwich ELISA. J Immunol Methods 167:227-236
White WL, Ferry JA, Harris NL, Grove AS Jr (1995) Ocular adnexal lymphoma. Ophthalmology 102:1994-2006
Whitnall SE (1912) The naso-lacrimal canal: the extent to which it is formed by the maxilla, and the influence of this upon its calibre. Ophthalmoscope 10:557-558
Wiede A, Hinz M, Canzler E, Franke K, Quednow C, Hoffmann W (2001) Synthesis and localization of the mucin-associated TFF-peptides in the human uterus. Cell Tissue Res303:109-115
Wiede A, Jagla W, Welte T, Köhnlein T, Busk H, Hoffmann W (1999) Localization of TFF3, a new mucus-associated peptide of the human respiratory tract. Am J Respir Crit Care Med 159:1330-1335
Wilhelmus KR. The importance of having lysozyme. Cornea 4:69-70
Williams R, Stamp GWH, Lalani E-N, Pignatelli M (1996) pS2 transfection of murine adenocarcinoma cell-line (410.4) enhances dispersed growth pattern in 3-D collagen gel. J Cell Sci 109:63-71

Wilson G, Merril R (1976) The lacrimal drainage system: Pressure changes in the canaliculus. Am J Optom Physiol Optics 53: 55-59
Wood JN, Anderton BH (1981) Monoclonal antibodies to mammalian neurofilaments. Bioscience Rep 1: 263-268
Wotherspoon AC, Hardman-Lea S, Isaacson PG (1994) Mucosa-associated lympoid tissue (MALT) in the human conjunctiva. J Pathol 174:33-37
Wotherspoon AC, Ortiz-Hidalgo C, Falzon MR, Isaacson PG (1991) Helicobacter pylori-associated gastritis and primary B-cell gastric lymphoma. Lancet 338: 1175-1176
Wright N (1998) Aspects of the biology of regeneration and repair in the human gastrointestinal tract. Phil Trans R Soc Lond B 353:925-933
Wright NA, Hoffmann W, Otto WR, Rio M-C, Thim L (1997) Rolling in the clover: the trefoil factor family (TFF)-domain peptides, cell migration and cancer. FEBS Lett 408:121-123
Yamagami S, Dana R (2001) The critical role of lymph nodes in corneal alloimmunization and graft rejection. Invest Ophthalmol Vis Sci 42:1293-1298
Yang D, Chen Q, Chertov O, Oppenheim JJ (2000) Human neutrophil defensins selectively chemoattract naive T and immature dendritic cells. J Leukoc Biol. 68:9-14
Yang D, Chertov O, Bykovskaia SN, Chen Q, Buffo MJ, Shogan J, Anderson M, Schroder JM, Wang JM, Howard OM, Oppenheim JJ (1999) Beta-defensins: linking innate and adaptive immunity through dendritic and T cell CCR6. Science 286:525-528
Yazici B, Hammad AM, Meyer DR (2001) Lacrimal sac dacryoliths – predictive factors and clinical characteristics. Ophthalmology 108:1308-1312
Zamboni L, de Martino C (1967) Buffered picric acid-formaldehyde: a new, rapid fixative for electron microscopy. J Cell Biol 35:148A
Zasloff M (2002) Antimicrobial peptides of multicellular organisms. Nature 415:389-395
Zolli CL, Shannon GM (1982) Dacryocystorhinostomy: a review of 119 cases. Ophthal Surg 13:905-910
Zucht HD, Grabowsky J, Schrader M, Liepke C, Jurgens M, Schulz-Knappe P, Forssmann WG (1998) Human beta-defensin-1: A urinary peptide present in variant molecular forms and its putative functional implication. Eur J Med Res 3:315-323

Subject Index

A
absorption 16, 46, 75
accessory lacrimal gland 1, 83
albumin 77
allergic conjunctivitis 72
animal model 78
antibiotic peptide 64
antimicrobial defence 64
antimicrobial peptide 3, 63, 66
antimicrobial peptides 7, 11
- BPI 15
- CAP-37 15
- human β-defensin 11
- human alpha defensin 5 15
- human alpha defensin 6 15
- human neutrophil defensin 11
- lactoferrin 11
- LL-37 15
- lysozyme 11
antimicrobial peptides, secretory phospholipase A_2 11
apoptosis 84
autoimmunity 85
autonomic control 74

C
canaliculi 66
canaliculus 1, 8
cavernous body 70
- barrier artery 70
- capacitance vein 70
- cushion vein 70
- Drosselvene 70
- Polstervene 70
- Sperrarterien 70
- Sperrvene 70
- throttle vein 70
cavernous body, arteriovenous anastomose 70
comparative anatomy 17, 49–50
- apes 8
- cat 8
- deer 8
- pig 8
- rabbit 8
- rat 8
conjunctiva 68, 83
cornea 83

D
dacryocystitis 63, 80
dacryolith 78
dacryostenosis 53, 66, 68, 80
defence cell 12, 33–34, 55
drainage 71
dry eye 4, 53, 82, 84
- keratoconjunctivitis sicca 4, 82
- sicca syndrome 4

F
feedback signal 82
"functional obstruction" 80
functional unit 83

G
germinal centre 67
goblet cell 59, 61–62, 77

H
^{3}H-cortisol 78
Horner's muscle 1

I
IgA 35–36
immune deviation 84
immune privilege 83
inflammation 81
inter-connecting neural-reflex arc 83

L
lacrimal fluid 75
lacrimal gland 1, 83
lacrimal passage 1, 4–5, 63, 68, 70–71, 75, 77, 80, 83
lacrimal portion of the orbicularis oculi muscle 1

lacrimal pump 75
lacrimal sac 1, 3, 8
lacrimal system 1, 3, 7, 63, 67, 75, 81
lectin 22
lipophilic substance 77
low-grade B-cell lymphoma 68
lymphoepithelium 67

M
M cell 68
mantle zone 67
marginal zone 67
Meibomian gland 83
metaplasia 81
mucin 2
mucins 59–62, 82
mucosa associated lymphoid tissue 31
- MALT 35
- TALT 38
mucosa-associated lymphoid tissue (MALT) 3
mucosal immune system 63
mucosal surface 59–60, 63, 66–67, 75
mucus 59–60, 63, 68

N
nasolacrimal duct 1, 3–5, 17, 21–22, 24–25, 27–29, 31, 33–34, 36, 38–40, 45, 47–54, 59, 61–62, 64, 66–67, 69, 71, 74, 77–78, 82
nervous tissue 12, 41, 44, 72
neurofilament 41
neuronal enzyme 41
neuropeptide 41, 44, 73

O
ocular surface 60, 64, 74, 78, 82, 84

P
Peyer's patches 68
primary acquired dacryostenosis 79
- functional 81
- idiopathic 79
- PANDO 79
primary acquired nasolacrimal duct obstruction (PANDO) 4

R
rabbit 77

S
secretory component 67
seromucous gland 72
Sjögrens syndrome 4
submucosal swelling 74

T
tear film 1, 4, 23, 27, 60, 78
- aqueous component 1
- lipid component 1
- mucous component 1
tear flow 3, 70, 74, 81
tear fluid 1, 3, 22, 41, 53, 69, 71, 75–77, 79, 82, 84
tear outflow 71
tears 84
TFF peptide 23, 27
TFF peptides 3, 61–62, 68
- trefoil factor peptides 11
TFF-peptides, TFF1 23
TFF-peptides, TFF2 23
TFF-peptides, TFF3 23
tolerance 83

V
vascular plexus 70
vascular system 37

GPSR Compliance

The European Union's (EU) General Product Safety Regulation (GPSR) is a set of rules that requires consumer products to be safe and our obligations to ensure this.

If you have any concerns about our products, you can contact us on ProductSafety@springernature.com

In case Publisher is established outside the EU, the EU authorized representative is:

Springer Nature Customer Service Center GmbH
Europaplatz 3
69115 Heidelberg, Germany

Batch number: 10704713

Printed by Printforce, the Netherlands